DRUNK

ALSO BY EDWARD SLINGERLAND

Trying Not to Try:
Ancient China, Modern Science, and the Power of Spontaneity

DRUNK

HOW WE SIPPED, DANCED, AND STUMBLED OUR WAY TO CIVILIZATION

Edward Slingerland

Little, Brown Spark

New York Boston London

Little, Brown Spark
Hachette Book Group
1290 Avenue of the Americas, New York, NY 10104
littlebrownspark.com

First Edition: June 2021

Little, Brown Spark is an imprint of Little, Brown and Company, a division of Hachette Book Group, Inc. The Little, Brown Spark name and logo are trademarks of Hachette Book Group, Inc.

The publisher is not responsible for websites (or their content) that are not owned by the publisher.

The Hachette Speakers Bureau provides a wide range of authors for speaking events. To find out more, go to hachettespeakersbureau.com or call (866) 376-6591.

ISBN 978-0-316-45338-7
LCCN 2020951436

Printing 5, 2021

LSC-H

Printed in the United States of America

T&T

BD, MD

AMFT

CONTENTS

INTRODUCTION 3

1. WHY DO WE GET DRUNK? 17

Brain Hijack: Porn and Sexually Starved Fruit Flies 23

Evolutionary Hangovers: Drunken Monkeys,
Liquid Kimchee, and Dirty Water 27

More Than Twinkies and Porn: Beyond Hangover
and Hijack Theories 31

A Genuine Evolutionary Puzzle: An Enemy in the
Mouth That Steals Away the Brain 35

A Genetic Mystery: We Are Apes Built to Get High 41

A Cultural Mystery: Prohibition's Strange Failure to
Take Over the World 49

Pickles for the Ancestors? 58

2. LEAVING THE DOOR OPEN FOR DIONYSUS 61

The Human Ecological Niche: Creative, Cultural,
Communal 65

The Creative Animal 71

The Cultural Animal 80

CONTENTS

The Communal Animal 87

Regaining the Child's Mind 95

The Drunken Mind 97

Leaving the Door Open for Dionysus 103

3. INTOXICATION, ECSTASY, AND THE ORIGINS OF CIVILIZATION 107

A Visit from the Muse: Intoxication and Creativity 111

Chemical Puppies: Turning Wolves into Labradors 116

The Chemical Handshake: *In Vino Veritas* 124

Puking and Bonding 134

Liquid Ecstasy and the Hive Mind 141

Political Power and Social Solidarity 150

Cultural Group Selection 156

4. INTOXICATION IN THE MODERN WORLD 159

Whiskey Rooms, Saloons, and the Ballmer Peak 162

Truth Is the Color Blue: Modern Shamans
and Microdosing 171

Why Skype Didn't Eliminate Business Travel 177

Office Parties: Pros and Not Just Cons 182

Long Live the Local 187

Beauty Is in the Eye of the Beer Holder: Sex,
Friendship, and Intimacy 192

Collective Effervescence: Tequila Shots
and Burning Man 199

CONTENTS

Ecstasy: Vacation from the Self   204

It's Only Rock-n-Roll: Defending the Hedonistic Body   211

It Is Time to Be Drunk   217

5. THE DARK SIDE OF DIONYSUS   223

The Puzzle of Alcoholism   226

The Problem with Liquor: An Evolutionary Mismatch   232

Isolation: The Danger of Drinking Alone   239

Distillation and Isolation: The Twin Banes of Modernity   244

Drunk Driving, Bar Fights, and Venereal Disease   247

Beer Goggles and Violence Against Women   251

Outsiders and Teetotalers Not Welcome:
Reinforcing the Old Boys' Clubs   254

Solace or Wedge? Reinforcing Bad Relationships   261

Drunk on Heaven: Getting Beyond Alcohol?   264

Taming Dionysus   270

Living with Dionysus   280

CONCLUSION   283

Acknowledgments   *293*

Bibliography   *297*

Notes   *323*

Index   *351*

This thirst for a kind of liquid which nature has sheathed in veils, this extraordinary need which acts on every race of mankind, in every climate and in every kind of human creature, is well worth the attention of the philosophical mind.

—Jean Anthelme Brillat-Savarin

DRUNK

INTRODUCTION

People like to masturbate. They also like to get drunk and eat Twinkies. Not typically all at the same time, but that's a matter of personal preference.

From a scientific perspective, we have long been told that these otherwise variegated pleasures have one thing in common: They are evolutionary mistakes, sneaky ways humans have figured out how to get something for nothing. Evolution gives us little shots of pleasure for doing things that advance its plan, like nourishing our bodies or passing on our genes. Clever primates, though, have been gaming this system for eons—inventing porn, birth control, and junk food, and seeking out or creating substances that will flood their brains with dopamine with callous disregard for evolution's original design goals. We are inveterate pleasure seekers, promiscuously grabbing little jolts of ecstasy whenever and wherever we can. When someone gets an endorphin hit from devouring a Twinkie, downing a shot of Jägermeister, and then pleasuring themselves to *Swingers Getaway IV,* they are getting an undeserved reward. Evolution must be furious.

One type of evolutionary mistake can be thought of as an evolutionary "hangover," where we are plagued by behaviors and drives that were once adaptive, but are no longer. Our desire for

Twinkies is a classic example of an evolutionary hangover. Junk food is appealing because evolution built us to like sugar and fat. This was a sensible strategy for our ancestors, hunter-gatherers haunted by the constant specter of hunger and starvation. It goes seriously off the rails, however, in modern environments, where most people have easy access to cheap sweets, carbs, and processed meats, sometimes helpfully delivered in a single, heart attack–inducing package. Evolution can also be subverted by "hijacks." These are cases where we've figured out an illicit way to tap into a pleasure system originally designed to reward other, more adaptive behavior. Masturbation is an exemplary hijack. Orgasms are meant to reward us for having reproductive sex, thereby helping our genes get into the next generation. We can, however, trick our bodies into giving us that same reward in any number of entirely, wildly non-reproductive ways.

In scientific circles, there is debate about whether our mistaken taste for alcohol is of the hijack or hangover variety. Proponents of hijack theories claim that alcoholic beverages make us feel good because their active ingredient, ethanol, happens to trigger the release of reward chemicals in our brain. This is a design glitch: These chemicals are actually intended by evolution to reward genuinely adaptive behavior, like eating nutritious things or pushing a hated enemy into a tar pit. But the brain can be tricked, and ethanol is one of the easiest ways to do so.

Proponents of the hangover theory see various ways in which a desire for getting at least mildly drunk might have been adaptive for our evolutionary ancestors, but argue that this drive has become extremely maladaptive in any kind of modern environment.

Whether of the hangover or hijack variety, evolutionary mistakes persist because natural selection hasn't bothered to deal with them yet. This is typically because whatever costs they involve are either relatively minor or have only become problematic quite recently.

Evolution can afford to turn a blind eye to masturbation as long as our drive for orgasms still results in enough genes getting passed on to the next generation. Junk food is a modern problem mostly confined to the developed world. Alcohol is also something evolution could afford to ignore, at least until relatively recently. This is because alcohol, like sugar, occurs only in small quantities in the natural world. It takes some serious work to get a buzz off naturally fermenting fruit. It is only with the advent of agriculture and organized, large-scale fermentation—maybe 9,000 years ago, a blink of an eye in evolutionary terms—that serious booze became available to lots of people, pushing susceptible humans onto the slippery slide to widespread drunkenness, lost weekends, and ruined livers.

A crucial but often unacknowledged feature of any sort of evolutionary mistake view of alcohol or other chemical intoxicant use is that it sees getting drunk or high, like masturbation or stuffing your face with junk food, as an unmitigated *vice*. A vice is a habitual practice that gives fleeting pleasure, but that is ultimately harmful to oneself and others, or at best a waste of time. Indeed, even the most ardent fan of masturbation would have to admit that, all else being equal, there are probably more productive ways to spend a weekend afternoon. Indulging in these practices may feel good, but it is not *doing* us—or anyone else—any good.

Not all vices are created equal, however. When it comes to our *Swingers Getaway IV* scenario, it's actually the Jäger shots that should keep evolution up at night. A bit of work time lost to masturbation is no big deal. Alcohol, on the other hand, can be truly dangerous. Alcoholic intoxication is an abnormal mental state, characterized by reduced self-control and degrees of either euphoria or depression, brought about by the temporary impairment of a big chunk of the brain. As the term suggests, it involves the ingestion of a *toxin*, a substance so harmful to the human body

5

that we possess elaborate, multi-layered physiological machinery dedicated to breaking it down and getting it out of our systems as quickly as possible. Our bodies, at least, clearly see alcohol as a serious threat.

An alcoholic beverage typically provides calories but little nutritional value, and is made from otherwise valuable, and historically scarce, grains or fruit. Its consumption impairs cognition and motor skills, damages the liver, kills off brain cells, and fuels ill-advised dancing, flirting, fighting, and even more louche behaviors. In small doses, it can make us happy and more sociable. But increased consumption quickly leads to slurred speech, violent arguments, maudlin expressions of love, inappropriate touching, or even karaoke. While getting completely wasted can induce ecstatic experiences of selflessness and group bonding, it also often leads to vomiting, injuries, blackouts, ill-advised tattoos, and serious property damage. And let's not even get started on hangovers.

From an evolutionary perspective, the use of certain drugs makes sense. Coffee, nicotine, and other stimulants are basically performance enhancers, allowing us to pursue our normal evolutionary goals with an extra spring in our step, our motor functions unimpaired, and our grip on reality firm.[1] It is the use of *intoxicants,* primarily alcohol, that is truly puzzling. This is because intoxicants, from the minute they hit our bloodstream, begin to impair us, slowing our reflexes, dulling our senses, and blurring our focus. They do this primarily by targeting the prefrontal cortex (PFC) in the brain, our center of cognitive control and goal-directed behavior. "Intoxication," as we'll be using the term in this book, thus includes not only the more dramatic states of inebriation— what we'd consider legally "drunk"—but also the mellow, happy buzz produced by those first couple sips of wine. As innocuous as a mild social high might seem, it is already undermining the capacity that arguably makes us human: our ability to consciously govern our own behavior, stay focused on task, and maintain a clear sense of self.

Given that the PFC is a key to our success as a species, consuming any amount of alcohol or other intoxicant seems really stupid. It takes well over twenty years to fully develop the PFC, a physiologically expensive part of the brain, and the last to reach maturity. It is therefore odd that one typical way to celebrate a twenty-first birthday is to chemically knock it down a few pegs. Given the potentially enormous costs, and apparent lack of benefits, to impairing our cognitive control, why do humans still like to get intoxicated? Why is the labor-intensive practice of converting wholesome grains and delicious fruit into bitter, low-dose neurotoxins, or seeking out intoxicating plants in the local biome, so ubiquitous across cultures and geographic regions?

It should puzzle us more than it does that one of the greatest foci of human ingenuity and concentrated effort over the past millennia has been the problem of how to get drunk. Even small-scale societies on the brink of starvation will set aside a good portion of their precious grain or fruit for alcohol production. In pre-colonial Mexico, tribes that otherwise had no organized agriculture traveled great distances to make liquor from cacti fruit during the brief periods when they were in season. Migrants whose alcohol supplies have run dry have desperately fermented shoe leather, grasses, local insects, whatever they could get their hands on. Nomads of Central Asia, with little access to starch or sugars, go so far as to make booze out of fermented mare's milk. In contemporary societies, people spend an alarming proportion of their household budgets on alcohol and other intoxicants. Even in legally dry nations, huge numbers of people suffer painful deaths trying to get drunk on cleaning products or perfumes.

The rare cultures that do not produce alcohol inevitably substitute some other intoxicating substance, such as kava, hallucinogen-laced tobacco, or cannabis, in its place. Among traditional societies, if there is something in the biome that has psychoactive properties, you can be sure that the locals have been using it for millennia. More often than not, it tastes horrible and has vicious side effects. For instance, ayahuasca, a hallucinogenic brew made from Amazonian vines, is painfully bitter and quickly brings on brutal diarrhea and vomiting. In some South American cultures, people go so far as to *lick poisonous toads.* All over the world, wherever you find people, you find them doing disgusting things, incurring incredible costs, and expending ridiculous amounts of resources and effort for the sole purpose of getting high.[2] Given how central the intoxication drive is to human existence, the archaeologist Patrick McGovern has only semi-facetiously suggested that our species be referred to as *Homo imbibens.*[3]

This desire to get mentally altered has ancient roots, ones that can be traced to the very beginnings of civilization.[4] At sites in eastern Turkey, dating to perhaps 12,000 years ago, the remains of what appear to be brewing vats, combined with images of festivals and dancing, suggest that people were gathering in groups, fermenting grain or grapes, playing music, and then getting truly hammered before we'd even figured out agriculture. In fact, archaeologists have begun to suggest that various forms of alcohol were not merely a by-product of the invention of agriculture, but actually a motivation for it—that the first farmers were driven by a desire for beer, not bread.[5] It is no accident that the earliest human archaeological finds from around the world always include huge numbers of specialized, elaborate vessels used solely for the production and consumption of beer and wine.

Sumerian myths go so far as to link the origins of human civilization to drinking (and good sex). In the epic *Gilgamesh* (ca. 2000 BCE), probably our oldest surviving literary document, the wild man Enkidu, who roams as one with the animals, is tamed and made human by a temple prostitute. Before offering him a full week of mind-blowing sex, she first sates him with the two great pillars of civilization: bread and beer. He particularly likes the beer, drinking seven jugs that cause him to "become expansive and sing with joy." Then, and only then, do they move on to the main event.[6] The ancient Aryans, who sometime between 1600 and 1200 BCE moved from the steppes of Central Asia into the Indian subcontinent, built their religious system around a mysterious intoxicant called "soma." Scholarly debate continues to rage about what soma actually was—the current dominant theory is that it was a liquid made from the fly agaric hallucinogenic mushroom[7]—but it clearly packed a punch. A hymn from the *Rig Veda*, dating to perhaps 1200 BCE, records the god Indra's words as the soma high starts to kick in and his thoughts begin to race, leaving him wildly out of his mind but also imbued with universe-shattering power:

The five tribes are no more to me than a mote in the eye.
　　Have I not drunk Soma?
The two world halves do not equal a single wing of mine.
　　Have I not drunk Soma?
In greatness, I surpass heaven and this great earth. Have I not
　　drunk Soma?
Yes, I will place the earth here, or perhaps there. Have I not
　　drunk Soma?
I will thrash the earth soundly, here or perhaps there. Have I
　　not drunk Soma?
One of my wings is in heaven, the other trails below. Have I
　　not drunk Soma?
I am huge, huge! Flying to the clouds. Have I not drunk
　　Soma?[8]

Why is one of the most important of Vedic gods imagined as not merely getting supremely lit up, but actually drawing his power from a concoction of magic mushrooms? This is particularly puzzling when the literal drug in question is more likely to leave one prostrate and helpless, pupils dilated and motor coordination shot, hardly in any kind of shape to "thrash the earth soundly." Would it not make more sense to portray Indra as having enjoyed a solid meal and nutritious milk before heading off to order the universe or dispatch his enemies?

The great power of adopting a scientific approach to human behavior is the ability to unmask deep puzzles about human existence that otherwise hide in plain sight. Once we begin to think deeply and systematically about the antiquity, ubiquity, and power of our taste for intoxicants, the standard stories suggesting it's some sort of evolutionary accident become difficult to take seriously. Considering the enormous costs of intoxication, which humans have been paying for many thousands of years, we would expect genetic evolution to work toward eliminating any accidental taste for alcohol from our motivational system as quickly as possible. If ethanol happens to pick our neurological pleasure lock, evolution should call in a locksmith. If our taste for drink is an evolutionary hangover, evolution should have long ago stocked up on aspirin. It hasn't, and explaining *why* it hasn't is of more than merely academic interest. Without understanding the evolutionary dynamics of intoxicant use, we cannot even begin to think clearly or effectively about the role intoxicants can and should play in our lives today.

While plenty of entertaining books have been written about the history of alcohol and other intoxicants, there has yet to be one that offers a comprehensive, convincing answer to the basic question of why we want to get high in the first place.[9] The sheer popularity, persistence, and importance of intoxicants throughout human history begs explanation. In the pages that follow, I aim to provide one. Cutting through the tangle of urban legends and anecdotal impressions that surround our notions of intoxication, I draw on evidence from archaeology, history, cognitive neuroscience, psychopharmacology, social psychology, literature, poetry, and genetics to provide a rigorous, scientifically grounded explanation for our drive to get drunk. My central argument is that getting drunk, high, or otherwise cognitively altered must have, over evolutionary time, helped individuals to survive and flourish, and cultures to endure and expand. When it comes to intoxication, the mistake story cannot be correct. There are very good evolutionary reasons why we get drunk.[10] What this means is that most of what we think we know about intoxication is wrong, incoherent, incomplete, or all of the above.

Let's begin with the first point. Evolution is not stupid, and works much faster than most people realize. Pastoralists have genetically adapted to drinking milk as adults, Tibetans to living at high elevations, and boat-dwelling Southeast Asian peoples to diving and holding their breath underwater within the space of a few generations.[11] If alcohol or drugs were merely hijacking pleasure centers in the brain, or were once adaptive millennia ago but are purely vices now, evolution should have figured this out pretty quickly and put a firm end to the nonsense. This is because, unlike porn or junk food, alcohol and other intoxicants are extremely costly, both physiologically and socially. Our genes face only a marginal cost when they allow us to waste a few moments masturbating or gain a few pounds eating Twinkies. Drunkenly plowing our car into a telephone pole, perishing from liver damage, or losing our livelihood and family to alcoholism are much more serious and direct threats to our genetic well-being. Similarly, cultures can afford to wink at harmless vices, especially ones that might keep people more docile and obedient. Marx never called pornography the opium of the people, but might have if he'd ever gotten a glimpse of the internet. *Literal* opium, though, is potentially terribly disruptive to cultures, as is any chemical intoxicant.

The fact that our supposedly accidental taste for intoxicants has not been eradicated by genetic or cultural evolution—even when perfectly good "solutions" exist, as I will explain below—means that something else must be going on. The cost of indulgence has to be balanced by specific, targeted benefits. This book argues that, far from being an evolutionary mistake, chemical intoxication helps solve a number of distinctively human challenges: enhancing creativity, alleviating stress, building trust, and pulling off the miracle of getting fiercely tribal primates to cooperate with strangers. The desire to get drunk, along with the individual and social benefits provided by drunkenness, played a crucial role in sparking the rise of the first large-scale societies. We could not have civilization without intoxication.

This leads to a second point. The fact that drinking facilitates social bonding may not sound like a world-shaking revelation. Without an understanding of the specific cooperation problems that confront humans in civilization, however, we have no way of explaining why, throughout history and across the world, alcohol and similar substances have been the almost universal go-to solution. Why bond over a toxic, organ-destroying, mind-numbing chemical when a rousing game of Parcheesi might suffice? Without an answer to *this* question, we have no way to intelligently weigh arguments for or against replacing after-work pub sessions with escape room competitions or laser tag outings. Many of us deliberately seek out a glass or two of wine to relax after a hard day at work. Would an afternoon bike ride work just as well? How about fifteen minutes of meditation? None of these questions can be answered without an understanding of the relevant biochemistry, genetics, and neuroscience.

Similarly, an ancient trope holds that poetic inspiration is found at the bottom of a bottle. Why is the bottle full of alcohol, not tea? What are the specific effects of alcohol consumption, how could it possibly help with creativity, and what is the right dosage for maximum effect? (Hint: well before you see the bottom of the bottle.) How does alcohol as muse compare to psilocybin or LSD, or simply a walk in the park? There are myriad puzzles surrounding intoxicant consumption that cry out for explanation, and there are currently no truly comprehensive ones on offer. Some people can (and do) drink like fish, others flush and get nauseous after a few sips of weak beer. Most people successfully integrate intoxicants into their daily lives, while others become dangerously addicted and disabled. What are the genes responsible for these reactions, and how can we explain their distribution across the world? All things considered, cultural norms forbidding intoxicant consumption seem like a pretty good idea. Why are they, in fact, relatively rare, and widely circumvented in practice? What are the implications for contemporary issues, such as the role of alcohol in the workplace and drinking age legislation? It should trouble us that our musings on such matters generally occur in complete ignorance of the relevant science. At best, we base our thinking on disconnected facts or snippets of scientific knowledge uninformed by a broader evolutionary perspective.

Although other forms of intoxication play a role in this story, there are good reasons for focusing primarily on alcohol in particular. Alcohol is the unchallenged king of intoxicants. It can be found almost anywhere people can. If you tasked a cultural engineering team with designing a substance that would satisfy specs aimed at maximizing individual creativity and group cooperation, they would come up with something very much like alcohol. A simple molecule. Easy to make out of almost any carbohydrate. Easy to consume. Storable. Precisely doseable.

Complex but predictable and moderate cognitive effects. Quickly eliminated from the body. Easily influenced by social norms. Can be packaged in a tasty delivery system. Pairs nicely with food. Whatever the benefits and functions of cannabis, soma, or dance-induced ecstasy, none of these intoxication technologies display this full range of features, and most also have significantly greater downsides. It's challenging to negotiate a treaty while high on mushrooms; the cognitive effects of cannabis show a high degree of variability between people; and dancing all night without food or sleep makes it really hard to show up for work in the morning. A two-cocktail hangover is, in contrast, a relatively minor burden to bear. This is why alcohol tends to displace other intoxicants when introduced into a new cultural environment, and has gradually become "the world's most popular drug."[12]

Chemical intoxication is clearly dangerous. Alcohol has ruined many lives, and continues to ravage individuals and communities across the globe. Beyond our vague cultural queasiness about celebrating pleasure for pleasure's sake, defending the benefits of alcohol risks provoking a strong backlash from those who rightly worry about the profound costs of intoxicant use. But understanding the evolutionary rationales for our drive to get high will help to inform conversations where we have hitherto—in our scientific and anthropological ignorance—been flying blind.

Our analysis will turn up some clear and easy-to-implement advice for everyday life, but also raise more complicated or contentious policy issues, such as the best role for alcohol in the workplace or university. In an age where we are growing rightly concerned about facilitating inappropriate behavior, we might very well decide that the answer is none, but this is not a foregone conclusion. We also have to reevaluate the historic benefits of intoxication, at both the individual and group level, in light of the unprecedented threats that intoxicants pose in the modern world. The relatively recent innovations of distillation and social isolation entirely change intoxicants' balance on the razor's edge between order and chaos, creating novel dangers that we only dimly appreciate.

To have survived this long, and remained so central to human social life, intoxication's advantages must have—over the course of human history—outweighed the more obvious negative consequences. What this calculus recommends in our modern world, massively complex and changing at an unprecedented rate, is something we can only evaluate properly when we take a broad historical, psychological, and evolutionary perspective. It's pretty clear that Twinkies are bad for you. Masturbation doesn't make you go blind, but has limited social benefits.

Making the case for alcohol is more complicated. Explaining the human thirst for intoxication is indeed, as the early modern French gastronome Brillat-Savarin put it, "well worth the attention of the philosophical mind." The answer to the question of why we get drunk—for what problems or challenges intoxicants provide a solution—is, however, of much more than merely philosophical or scientific interest. Understanding the functional role of our drive to get drunk will give us a better sense of the proper role of alcohol and other intoxicants in our lives today. Given the potential costs of getting it wrong, the stakes are too high for us to stumble along as we have, guided only by folk notions, dimly understood policies, or Puritanical prejudices. History can tell us *when* and *with what* we have gotten drunk. But it is only when we couple history with science that we can finally begin to understand not only *why* we desire to get drunk in the first place, but also how it might actually be good for us to tie one on now and then.

CHAPTER ONE

WHY DO WE GET DRUNK?

People love to drink. As the anthropologist Michael Dietler notes, "Alcohol is by far the most widely and abundantly consumed psychoactive agent in the world. Current estimates place the number of active consumers at over 2.4 billion people worldwide (or roughly one third of the Earth's population)."[1] And this is not a recent development: humans have been getting drunk for a really long time.[2] Images of imbibing and partying dominate the early archaeological record as much as they do twenty-first-century Instagram. A 20,000-year-old carving from southwestern France, for instance, shows a woman, possibly a fertility goddess, holding a horn to her mouth. One might imagine that she is using it as a musical instrument, blowing into it to produce sound, except for the fact that the part near her mouth is the wide end. She is *drinking* something, and it is hard to imagine that it's just water.[3]

The earliest direct evidence of alcoholic beverages deliberately being produced by human beings dates from around 7000 BCE in the Yellow River Valley of China, where potsherds from an early Neolithic village were found to contain chemical traces of a sort of wine, probably not very pleasant by modern standards, made from wild grapes and other fruits, rice, and honey.[4] There is evidence in present-day Georgia of grape domestication from circa 7000 to

6000 BCE. Pottery fragments from the same region, depicting human figures throwing their arms in the air in celebration, suggest that these grapes were intended for the cup and not the table.[5] Chemical evidence of grape wines, preserved with pine resin (as is still the case with Greek and other wines), has been found in present-day Iran in ceramics dating back to 5000 to 5500 BCE, and by 4000 BCE wine production had become a major collective undertaking. A huge cave site in Armenia apparently served as an ancient, full-scale winery, with basins for grape-stomping and pressing, fermentation vats, storage jars and drinking vessels.[6]

Figure 1.1. "Venus with a horn from Laussel" (Collection Musée d'Aquitaine; VCG Wilson / Corbis via Getty Images).

Neolithic peoples were also creative in terms of what they threw into their booze: In the Orkney Islands in northern Britain, archaeologists have discovered enormous pottery jars dating back to the Neolithic era that appear to have contained alcohol made from oats and barley, with the addition of various flavorings and mild hallucinogens.[7] The human drive to produce alcohol is impressive in its inventiveness as well as its antiquity. Inhabitants of Tasmania would tap a species of gum tree, dig a hole at its base, and allow the accumulated sap to ferment into an alcoholic beverage; the Koori people of what is now Victoria in southeast Australia fermented a mixture of flowers, honey, and gum into an intoxicating liquor.[8]

As the existence of ancient hallucinogenic beers suggests, although alcohol has remained the drug of choice among most large world cultures, humans have been wildly promiscuous when it comes to choosing their poison, supplementing alcohol with other intoxicating substances or finding replacements in places without alcohol.[9] Hallucinogens, typically derived from vines, mushrooms, or cacti, are a favorite, and sometimes given a special status above alcohol. The Vedic people of ancient India, for instance, possessed alcohol, but were a bit suspicious of it, thinking it produced a morally questionable form of intoxication. First in cultural and religious prestige was the psychological state, mada, produced by the hallucinogenic drug soma. "Mada" comes from the same root as the English word "madness," but in Sanskrit means something more like rapture or bliss, a privileged state of religious ecstasy.

Peyote buttons and mescaline-containing beans carbon-dated to 3700 BCE have been found in human cave dwellings in northeast Mexico.[10] Enormous stone carvings with human faces or animals incorporated into images of psilocybin mushrooms, and ceramics depicting mescaline cacti atop shamanistic animals, like jaguars, date back as far as 3000 BCE, suggesting that hallucinogens have long played a central role in religious rituals throughout Central and South America.[11] Over a hundred species of hallucinogens are found in the New World, and all have been intensively utilized by humans for millennia. The oddest hallucinogen has got to be the skin secretion of certain poisonous toads found in Central America, which can be enjoyed by drying the skin and smoking it or adding it to liquid concoctions.[12] Or, if you're in a hurry, you can also simply pin the toad down and *lick* it.

In the Pacific, cultures that never adopted alcohol use—possibly because alcohol would interact negatively with toxins acquired by consuming local seafood—ended up turning to kava as their preferred intoxicant.[13] Made from the root of an intensively domesticated crop, possibly first brought under human control in the island of Vanuatu, kava has been cultivated by humans for so long that it can no longer reproduce on its own.[14] It has both narcotic and hypnotic effects, and is a powerful muscle relaxant. Traditionally chewed up and spit into a bowl, which is then passed around in a manner strictly regulated by ritual, kava induces a contented and sociable state of mind, providing a more mellow high than alcohol.

And, speaking of mellow, we would be remiss in failing to mention cannabis, which is native to central Asia. Humans in Eurasia appear to have been lighting up and tuning out for at least 8,000 years, with cannabis becoming a widely traded and consumed ritual and recreational drug by 2000 BCE.[15] To get a sense of how old our fondness for pot is, one need look no further than a burial site in Central Eurasia, dating to the first millennium BCE, where a male tomb occupant was found draped in a shroud made of over a dozen cannabis plants.[16] In the fifth century BCE, the Greek historian Herodotus described terrifying Scythian warriors—horse-mounted nomads from Central Asia—chilling out by erecting wooden-frame tents, setting an enormous bronze stove in the center, throwing in a generous handful of cannabis, and proceeding to get wildly high. This practice has been confirmed by recent archaeological evidence, and it is thought that the Central Asian tradition of sparking up goes back five or six thousand years.[17] The Dude would be proud.

People outside Eurasia, without access to cannabis, made do with other smokes and chews. For millennia, natives of Australia have produced a mixture of narcotics, stimulants, and wood ash, called "pituri," and used it like chewing tobacco, holding a wad in their cheeks. The active ingredients are various strains of native tobacco and a local narcotic shrub (often also referred to as "pituri"). It is significant that in North America, one of the few places on the globe where native populations did not produce and use alcohol, there existed instead a highly elaborate system of tobacco cultivation and regional trade, with archaeologically recovered pipes dating back to somewhere between 3000 and 1000 BCE.[18] Although we do not tend to think of tobacco as an intoxicant, the strains cultivated by Native Americans were much more powerful and intoxicating than what you can now buy at your corner store. When mixed with hallucinogenic ingredients, as it typically was, it really packed a punch.[19] Opium is another drug that has been enjoyed by humans since our distant ancestors first figured out what it could do to their brains. Remains in Britain and Europe suggest that people were consuming opium poppies as long as 30,000 years ago,[20] and archaeological evidence shows that poppy goddesses were worshipped in the Mediterranean as far back as the second millennium BCE.[21]

So, people have been getting intoxicated—drunk, stoned, or lit up with psychedelics—for a really long time, all over the world. There is no shortage of entertaining books documenting our species' taste for intoxicants, as well as the wildly diverse ways in which we have pursued our desire for altered states.[22] As the alternative medicine guru Andrew Weil observes, "The ubiquity of drug use is so striking that it must represent a basic human appetite."[23] In his overview of the impressive variety of intoxication technologies used around the world, the archaeologist Andrew Sherratt similarly argues that "the deliberate seeking of psychoactive experience is likely to be at least as old as anatomically (and behaviorally) modern humans: one of the characteristics of *Homo sapiens sapiens.*"[24]

Typically left unexamined in these historical and anthropological surveys of our taste for booze, however, is the fundamental puzzle of *why* humans want to get drunk in the first place.[25] Practically speaking, getting drunk or high seems like a really bad idea. At the individual level, alcohol is a neurotoxin that impairs our cognition and motor function and damages our body. At the social level, the link between drunkenness and social disorder is not an invention of modern football hooligans or college students. Wild, dangerously chaotic bacchanalia—a word derived from the name of the Greek god alternately called Dionysus or Bacchus—were a standard feature of ancient Greek life. Descriptions and visual depictions of alcohol-fueled rituals and banquets from ancient Egypt to China make it clear that disorder, fighting, illness, poorly timed unconsciousness, copious vomiting, and illicit sex have long been common outcomes of alcohol consumption.

The various hallucinogens used by humans around the world are even more dangerous and disruptive. Besides completely disconnecting you from reality, their very chemical makeup can easily get you killed. A small shrub that grows in the Sonoran Desert, *Sophora secundiflora,* produces a bean that is so toxic that a single one will almost instantly kill a child. You would think people would very quickly learn to keep their distance. They haven't. This is because the so-called "mescal bean" can also get you super high. Although it has no known culinary value, traces of the bean have been discovered in archaeological remains dating back into the deep millennia BCE, when desert cultures were clearly using it for its intoxicating power. Half a bean is the proper dose for an adult, but you don't want to get it wrong. Eating more will produce "nausea, vomiting, headache, sweating, salivation, diarrhea, convulsions, and paralysis of the respiratory muscles. Death is by asphyxiation."[26] No doubt there were quite a few casualties before people finally worked this out.

Why take the chance? Whether we are talking about terrifyingly dangerous hallucinogenic beans or stupefying narcotics or disorienting, toxic alcohol, why do people not just say no? Given the costs and potential harm of intoxicants, we are justified in dismissing weak, ad-hoc justifications, like old wives' tales about aiding digestion or warming the blood. A prohibition crusader in the early nineteenth century rightly mocked the kind of evidence-free rationalizations that people are wont to spout in order to justify hitting the bottle:

> Strong drink in some form is the remedy for every sickness, the cordial for every sorrow. It must grace the festivity of the wedding; it must enliven the gloom of the funeral. It must cheer the intercourse of friends and enlighten the fatigue of labor. Success *deserves* a treat and disappointment *needs* it. The busy drink because they are busy; the idle because they have nothing else to do. The farmer must drink because his work is hard; the mechanic because his employment is sedentary and dull. It is warm, men drink to be cool; it is cool, they drink to be warm.[27]

We can do better than this. Let's begin by looking at the standard scientific explanations for the human drive to get drunk. They look better, at first glance, than the rationalizations mocked by prohibitionists, but in the end they are similarly unsatisfying.

BRAIN HIJACK: PORN AND SEXUALLY STARVED FRUIT FLIES

People like orgasms. From a scientific perspective, this is not puzzling. Orgasms are pleasurable because they are evolution's way of telling us: "Nice job. Keep doing what you were just doing." Evolution does this because, in the environments in which we evolved, an orgasm is a sign that you have been striving toward its central goal, which is getting your genes into the next generation.

It's not a perfect system, to be sure. All sorts of animal species have been gaming it since it was invented, from masturbating monkeys to leg-humping dogs. Humans are the worst, though. For instance, *Homo sapiens* have been producing pornography for about as long as they have been doing anything at all. It seems that any new technology—stone carving, painting, lithographs, cinematography, the internet—is, at first, used primarily for pornography. The sorts of voluptuous figures that pop up in prehistoric archaeological sites, like the Venus figure pictured earlier, are typically glossed by scholars as fertility or mother goddesses. Maybe. It's equally likely that they are early precursors to Playboy centerfolds and served the same purpose for the people who made them. In any case, from ancient erotica to modern sex dolls, when it comes to tricking evolution, on this as on most fronts, humans are really unmatched.

Evolution, though, has remained rather blasé about this chicanery. It doesn't care about perfection; it's happy with good enough. In the absence of reliable birth control, the basic design linking orgasms to the good work of passing on one's genes has historically functioned pretty well. Recent technological developments, however, seriously disrupt this link. Condoms and birth control pills effectively divorce the act of sex from the outcome it was designed to produce. The printing press, glossy magazines, VHS tapes, DVDs, and finally the internet provide a previously unimaginable quantity and variety of sexual images to any individual in the privacy of his or her own home. This concerted hijacking of our reward systems, taken to these extremes, may indeed partially undermine evolution's plans.

Perhaps the most common view of our taste for intoxication is that it involves precisely this sort of hijacking of previously adaptive drives. "Hijack" theories see alcohol and other intoxicating drugs as, like pornography, something that just happens to trigger reward systems in our brains originally designed by evolution to encourage adaptive behavior like sex. This was not a problem for most of our evolutionary history, when such drugs were hard to get in any quantity and were relatively weak in potency. Evolution could afford to ignore the fact that primates and other mammals enjoyed an occasional high from some fermented fruit found on the jungle floor in the same way it could overlook a bit of masturbation or non-reproductive sex. It could not, however, anticipate that one of these primates, with its big brain, tool use, and ability to accumulate cultural innovations, would suddenly—in an evolutionary blink of an eye—figure out how to make beer, wine, and then mind-bogglingly powerful distilled spirits. Hijack theories claim that these poisons have been able to slip through our evolutionary defenses because evolution is a sluggard in the face of rapid human innovation.

A classic exponent of this view is the founder of the field of evolutionary medicine, Randolph Nesse, who writes:

> Pure psychoactive drugs and direct routes of administration are evolutionarily novel features of our environment. They are inherently pathogenic because they bypass adaptive information processing systems and act directly on ancient brain mechanisms that control emotion and behavior. Drugs that induce positive emotions give a false signal of a fitness benefit. This signal hijacks incentive mechanisms of "liking" and "wanting," and can result in continued use of drugs that no longer bring pleasure...Drugs of abuse create a signal in the brain that indicates, falsely, the arrival of a huge fitness benefit.[28]

The evolutionary psychologist Steven Pinker similarly sees our modern use of intoxicants as a result of the confluence of two features of the human mind: our liking for chemical rewards and our ability to problem solve. A substance that manages to pick the pleasure lock in our brain, however accidentally, is going to become a focus of our goal-seeking and innovation, even if the pursuit of this substance has—from a purely adaptive perspective—neutral or negative consequences.[29] Our sex drive, as we've noted, is another good example of this dynamic. Evolution gives us a powerful incentive system, in the form of sexual pleasure and orgasms, and then dusts its hands and walks away contentedly, foolishly thinking it's just ensured that we will now exclusively pursue heterosexual, vaginal intercourse and thereby get our genes into the next generation. It clearly has no idea what humans are capable of. As an example of a maladaptation caused by the hijacking of reward systems, Pinker notes that "people watch pornography when they could be seeking a mate." Of course, this is only one thread of the rich tapestry of non-reproductive sexual hijinks to which we are prone, but it suggests why evolution should be keenly alert to the subversion of its designs.

A study that involved taunting sexually deprived fruit flies reinforces this concern. Given how tiny and apparently utterly unlike us they are, fruit flies (*Drosophila*) are surprisingly good proxies for humans in many respects, including the way they process alcohol.[30] Fruit flies like booze, they get drunk, and it stimulates their reward systems in a manner similar to ours. They can also become alcoholics: flies come to prefer heavily alcohol-laden food over regular food, and this desire becomes more powerful over time. If deprived of alcohol, they go on binges when it is reintroduced.[31] All of this is clearly maladaptive, at least at the levels of alcohol used in the lab, where the spiked food is often brought to the strength of a head-banging Australian Shiraz (about 15 to 16 percent alcohol).

Shiraz-drinking fruit flies have trouble flying straight, and therefore locating food and mates. The study with the sexually deprived fruit flies further found that, in essence, when denied sex they turn to the bottle.[32] Alcohol consumption artificially triggers the same reward signal as successful mating, which means that drunk fruit flies have a reduced interest in courtship behavior, since they are getting their pleasure elsewhere. Fine for the fruit flies, maybe, but not so great for their genes.[33]

EVOLUTIONARY HANGOVERS: DRUNKEN MONKEYS, LIQUID KIMCHEE, AND DIRTY WATER

Hijack theories overlap somewhat with the hangover theories described in the Introduction, which see our taste for intoxicants as a novel evolutionary problem. Hangover theories, however, see certain features of human psychology not as a purely accidental hijacking of our reward systems, but rather as something that originally did serve a good adaptive purpose, albeit one that has outlived its usefulness. Junk food is a classic example. Evolution has designed us to receive little shots of reward for consuming high-density caloric packages, especially if they contain fat or sugar. Being blind and rather slow-moving, it could not have anticipated the advent of convenience stores offering cheap, endless quantities of processed, sugary treats, potato chips, and processed meat products.

When it comes to explaining our taste for alcohol, perhaps the most prominent hangover theory is the "drunken monkey" hypothesis advanced by the biologist Robert Dudley.[34] In the tangled rain forests where humans first evolved, alcohol is produced in ripe fruit by yeast cells as part of their chemical warfare campaign against bacteria, which are less tolerant of alcohol and compete with the yeast for the fruit's nutrients. Alcohol therefore owes its very existence to a vicious history of yeast-bacteria warfare. Dudley argues that an incidental feature of the molecule that we call alcohol (technically, ethanol) is the key to why primates acquired a taste for it. Ethanol is extremely volatile—that is, it is a small, light molecule that can travel long distances in the air. It is therefore ideally positioned to function as an olfactory dinner gong for a wide variety of species. This no doubt includes fruit flies, whose taste for alcohol is probably linked to its ability to lead them to fruit.

It is Dudley's contention that this was also the case for early humans, as well as our primate ancestors and cousins, who—following the waft of alcohol molecules to find and identify the rare prize of ripe fruit—came to associate small amounts of alcohol with high-quality nourishment. Individuals who were particularly enamored of its taste or pharmacological effects would have been more likely to seek it out, acquiring more calories than their tee-totaler compatriots. This adaptive advantage favored the evolution of a taste for alcohol, as well as an ability to metabolize it. So, Dudley's argument is that alcohol makes us feel good because, in our evolutionary environment, it led to a large caloric and nutritional payoff. It's simply an evolutionary hangover that modern urbanites still derive pleasure from alcohol when it now only tends to lead to liver damage, obesity, and premature death. As Dudley puts it, "What once worked safely and well in the jungle, when fruits contained only small amounts of alcohol, can be dangerous when we forage in the supermarket for beer, wine, and distilled spirits."[35]

Other hangover theories argue that the fermenting of grains and fruits plays a useful role in converting their calories into a more durable, portable form, allowing the preservation of resources that would otherwise be lost in a world without refrigerators.[36] Alcohol, in this view, traditionally functioned like a more fun version of kimchee or pickles. This is clearly not an insignificant benefit of fermentation: even today, entrepreneurs in northern Tanzania ferment banana- and pineapple-based wines to preserve fruit that would otherwise quickly rot after harvest—and, of course, to produce a tasty brew.[37] Another advantage of fermentation, at least when we're talking about the transformation of grains into beer, is what the British nutritionist B. S. Platt, observing that fermenting maize into beer nearly doubles its essential micronutrient and vitamin content, called "biological ennoblement."[38] This nutritional transformation, caused by the action of yeast on fermenting grain, could have been particularly important in pre-modern agricultural societies. The archaeologist Adelheid Otto argues that, at least in Mesopotamia, the nutritional content of beer played a crucial role in rounding out people's "depressingly bad diet," which otherwise consisted almost entirely of starches and precious little fresh vegetables, fruit, or meat.[39] Even as recently as pre-Victorian England, it is thought that beer made up a significant portion of the average person's caloric intake.[40]

This points to another advantage of alcohol for pre-modern people: its simple caloric punch. A gram of pure alcohol packs 7 calories, compared to 9 calories for fat and 4 for protein. It is disturbing to note that a modest 5-ounce pour of red wine contains as many calories as a 2-inch square of brownie or small scoop of ice cream (around 130 calories). Studies have estimated that in certain historic and even contemporary cultures, beer can constitute up to one-third or more of local caloric intake.[41] As is depressingly familiar to anyone on a diet, alcoholic beverages are so calorically dense that there is a bit of truth in the tagline of that venerable stout, Guinness: "a meal in every glass." As with many aspects of our biology, what is a problem for modern tipplers could have been a great benefit to our chronically hungry, nutritionally stressed ancestors.

Another category of hangover theories focuses not on alcohol's volatility, or ability to preserve calories or add vitamins, but on its germ-killing properties. As we've noted, alcohol is designed to kill bacteria, having been produced by yeast as a weapon in their competition with bacteria to gain the upper hand in decomposing fruit and grain. This is why pure alcohol is an excellent disinfectant. Even in the weaker forms typically consumed by humans, it appears to retain some anti-microbial and anti-parasitic properties. This is why it's not a bad idea to drink when you have sushi: Washing down the raw fish with sake might help kill any nasty bugs that came along for the ride.

Even fruit flies take advantage of alcohol in this way. We've noted that they can be impressive tipplers, and their fruit-based diet makes them—like yeast—relatively tolerant of alcohol. A very cool evolutionary trick is performed by fruit flies when they sense the presence of parasitic wasps. These wasps are nasty predators that rather unkindly deposit their own eggs inside those of the fruit fly. Under normal conditions, this egg develops into a small wasp larva, which then feeds off the fruit fly larvae, completely devouring them from the inside before emerging to seek out new victims. In an environment where such wasps are a threat, female fruit flies seek out fruit with a high alcohol content on which to lay their eggs. Alcohol is not great for their own larvae, slowing their growth, but little fruit flies tolerate ethanol much better than the sensitive wasp larvae, which are generally killed off. Sacrificing some proportion of their offspring to alcohol poisoning is a small price to pay if at least some will survive. The fruit fly's relative tolerance for alcohol, initially driven by its reliance on fruit as food source, is in this way effectively weaponized and turned against a hated adversary.[42]

Finally, the process of fermenting alcoholic beverages also has the effect of disinfecting the water from which they are made. For much of human history, especially after the advent of agriculture and dense urban living, local water sources have often been extremely unsafe for drinking. It is possible that alcoholic fermentation therefore played a role in converting contaminated water into potable liquids. In some South American communities, *chicha,* a corn-based beer, remains an important source of hydration in regions that lack water treatment.[43] Medicinal properties have similarly been cited to explain our taste for plant-based intoxicants, many of which—in addition to causing us to see strange colored shapes, gods, or talking animals—are reasonably potent anti-parasite medicines.[44]

MORE THAN TWINKIES AND PORN: BEYOND HANGOVER AND HIJACK THEORIES

To the extent that people have seriously questioned the origins of our taste for intoxication, very few get beyond these kinds of "Twinkies and porn" accounts. They are, on the face of it, not implausible. Hangover theories, in particular, are intuitively compelling because there is probably some truth to them: Alcohol really does perform all of these super-useful functions. Its smell *can* signal high-payoff fruit rewards. It has nutritional value, it disinfects, and it sure does taste good.

At the end of the day, though, they all leave one feeling unsatisfied, like a half a pint of lukewarm near-beer on a hot, sunny afternoon. Hijack theories run up against the solid brick wall of the sheer, brutal cost of alcohol and other intoxicant consumption. Hangover theories like the drunken monkey hypothesis have received a lukewarm reception among primatologists and human ecologists, who point out that wild primates appear to avoid the kind of overripe fruit that produces ethanol, and that studies with humans suggest we strongly prefer simply ripe (no ethanol) fruit to overripe fruit.[45] (I certainly do.) Other hangover theories are hampered by the unfortunate fact that the postulated functions of alcohol or other drugs in our ancestral environment could have been performed just as well by something that doesn't paralyze large portions of your brain and leave you with a splitting headache in the morning.

For example, biologically "ennobling" a grain such as wheat, millet, or oats can be done simply by fermenting it into a porridge, as is still a common practice in small-scale agricultural communities around the world. Fermented porridges also solve the storage problem. A traditional Irish practice, for instance, is to turn oats into a porridge that ferments for weeks, gradually solidifying into a bread-like mass that can be sliced and fried up when needed. Delicious, especially if combined with bacon. Turning grains into porridges makes more efficient nutritional use of them than brewing them into beer. Oat porridge won't get you pleasantly buzzed, to be sure, but that raises the question of why we are vulnerable to such brain hijacking in the first place. If food preservation is the driving consideration, why didn't evolution select for individuals who went wild for porridge instead of beer? They would be presumably healthier and more productive than their beer-drinking cousins, and a culture that stuck to porridges alone would avoid a lot of reckless behavior, physical accidents, bad singing, and hangovers. From what we know, however, soothing breakfast porridges have historically served in Ireland as comfort the morning after, rather than a substitute for the substances that caused the discomfort in the first place.

Or consider the dirty water hypothesis. If you are suffering from bacteria-laden water, just *boil* it. Of course, the germ theory of disease is quite recent, and there are still people around the world who haven't gotten the news. But as human solutions to most adaptive problems have shown, we don't need to know anything about the actual causality involved to solve a problem through trial and error. Individuals do this all the time. Cultures are even better at it, since they can "remember" particularly good, chance solutions to problems and pass them on, benefiting the individuals in the cultures and/or aiding the spread of the group itself.[46]

Imagine a scenario where multiple groups are in competition for resources in a landscape full of rivers and lakes, but plagued by a high load of water-borne pathogens. We don't have to worry about groups that don't make alcohol, since they died off a long time ago—not surprisingly (to outside observers) around the time the water started going bad. The surviving groups have discovered alcohol, and developed the practice of only drinking beers that have effectively purified the water through fermentation. One group, though, finds that drinking the water in which they've boiled their evening meal of fish leaves them feeling a bit more sprightly in the morning, with a touch less diarrhea, stomach cramping, and other symptoms with which those of us who have drunk water we shouldn't have are all too familiar. A few individuals start drinking only this magic "fish water," eschewing beer and untreated water. They become more active, healthy, and successful than those who don't, so gradually this group comes to believe that only water blessed by the Fish God is fit for human consumption, and that all other beverages are taboo. The Clan of the Fish God begins to outcompete their beer-drinking neighbors. The beer-drinkers are similarly free of water-borne disease, but the hangovers and fatigue caused by late-night drinking sessions mean they're a little slower to the fishing grounds in the morning. The fish-water drinkers gradually begin to eliminate or assimilate the beer-drinkers; or the beer-drinking groups see the light and decide to convert to the cult of the Fish God and renounce all other beverages. Within a few generations of the discovery of fish water, alcohol use has been completely wiped out.

Perhaps the most obvious piece of cultural-historical evidence against the idea that the need for water purification drove the invention of alcohol is the case of China. People in the Chinese cultural sphere have been drinking tea forever (well, for at least a few thousand years), and for a long while have also had powerful cultural norms against drinking untreated water. Of course, that's not how they frame it: According to traditional Chinese medical beliefs, drinking cold water harms the *qi*, or energy, of the stomach. If you must drink water, it should be "opened water" (*kaishui*), boiled and drunk warm or at least at room temperature. The theory focuses on temperature and its effect on qi, not on the danger of water-borne pathogens, but it has the same function: Don't drink water unless it's been boiled and the nasty stuff has been killed off. It seems, then, that Chinese and Chinese-influenced cultures, which together encompass a pretty impressive proportion of the people who have ever lived on the earth, have solved the pathogen-load problem through the simple expedient of drinking only tea or boiled water.

And yet they still have booze. Oceans of it. From ancient Shang times (1600 to 1046 BCE) to the present, alcohol has dominated ritual and social gatherings in the Chinese cultural sphere as much as, if not more than, anywhere else in the world. This makes no sense if killing off pathogens in our water or stomachs was the main function of alcoholic beverages. Once the Chinese discovered tea and adopted norms against drinking untreated water, alcohol use should have tailed off and then disappeared, its primary function having been taken over by something much less dangerous, costly, and physiologically harmful. The unfortunate continued existence of *baijiu* ("white alcohol"), a brutally effective sorghum-based spirit, reminds us that this is not the case. It is also important to note that the dirty water hypothesis does not, in fact, fit with other cultural norms we see when we look around the world. Groups that have beer or wine typically still drink untreated water, or mix untreated water into their booze.[47] None of this makes sense if the main adaptive function of alcohol is to help us avoid bad stomachs.

Given the obvious costs of consuming alcohol, then, cultural evolutionary dynamics suggest that alternate solutions to the problems of dirty water, lack of micronutrients, or food preservation would be quickly discovered and exploited, driving alcohol use into extinction. This has not happened, to say the least.

A GENUINE EVOLUTIONARY PUZZLE: AN ENEMY IN THE MOUTH THAT STEALS AWAY THE BRAIN

Whether framed in terms of brain hijacking or evolutionary hangovers, existing theories all agree in seeing our taste for intoxication as a mistake, and in arguing that there is little or no functional role for intoxicants in contemporary human societies. Need to locate calorie-dense regions in your environment? Go to a supermarket. Need to preserve your food? Put it in a refrigerator. Have a problem with worms in your stool? Most doctors would recommend a prescription anthelmintic over a pack of cigarettes. Dirty water? Just boil it. Yet the fact remains that people still really like to drink and get high, resisting what would seem like strong selections pressures to the contrary. Cultural groups have been similarly stubborn in their dogged enthusiasm for alcohol and other drugs.

What is wonderful about evolutionary approaches is not only that they help us to explain otherwise puzzling aspects of human behavior, they also allow us to recognize the existence of these puzzles in the first place. Take, for instance, religion. I was trained as a historian of religion, and my entire field has always simply taken for granted—as a basic, unremarked-upon starting point—

the fact that human beings, across the world and throughout time, have believed in invisible supernatural beings, sacrificed enormous amounts of wealth to them, and incurred great costs to serve them. The laundry list of painful, costly, or merely super-inconvenient behaviors inspired by the world's religions is startling, once you start to think about it. Cutting off the foreskin of your penis, forgoing delicious and nutritious shellfish and pork, fasting, kneeling, self-flagellation, chanting mantras, sitting through hours of boring sermons in an uncomfortable suit on your only day off, jabbing metal stakes through your cheeks, stopping everything you are doing to bow in a particular direction five times a day. None of this makes any biological sense. It is when we put on our Darwinian spectacles that the baffling nature of this behavior becomes apparent.

Humans in groups are similarly profligate in their modes of worship. In ancient China, a good chunk of GNP was simply buried in the ground with dead people. Visitors to the tomb of the First Emperor of Qin marvel at the detail of the individual terra-cotta soldiers, the fully intact chariots, the awesome spectacle of a full army arrayed to protect the dead emperor. Rarely, if ever, does the question arise of why anyone would build something so monstrously wasteful in the first place. Remember, all of this was built at enormous expense and then simply *buried in the ground,* along with a disturbing number of freshly sacrificed horses and people. And China is no outlier. Think of Egyptian or Aztec pyramids, Greek temples, Christian cathedrals. It's a good bet that the largest, most expensive and lavish structure in any pre-modern culture is dedicated to religious purposes.

Looked at from an evolutionary perspective, this is all really stupid. Assuming, as we do as scientists, that the supernatural beings supposedly being served do not, in fact, exist, religious behavior is profoundly wasteful and counter-adaptive. Since no supernatural punishment is forthcoming, an individual who forgoes the pain and danger of sticking metal stakes through his cheeks, spends his time pursuing pragmatic aims rather than praying to a non-existent being, and enjoys protein and calories wherever they can be found should be more successful, healthier, and therefore leave more descendants than a religiously observant person. Since nonexistent ancestor spirits have no power to punish the living, cultures that invested their labor in improving city walls, building irrigational canals, or training their armies, rather than erecting useless monuments or burying entire fake armies in the ground, should have outcompeted religious groups. And yet this is not what we see in the historical record. The cultures that survive and expand and gobble up other cultures tend to go in for waste and human sacrifice on a grotesque scale. As scientists, we can only conclude that some other adaptive forces must be at work, such as the need for group identity or social cohesion.[48]

The use of intoxicants should puzzle us as much as religion does, and is similarly ripe for a proper scientific examination. Yet, as in the case of religious belief and practice, the very ubiquity of human intoxication renders the mystery of its existence invisible. It is only when we look at intoxicant use through the lens of evolutionary thinking that the truly odd nature of the phenomenon becomes clear. Given the social costs of alcohol and other chemical intoxicants—domestic abuse, drunken brawling, wasted resources, hungover and useless soldiers and workers—why has the production and consumption of alcohol and similar substances remained at the heart of human social life? George Washington famously prevailed over a much superior force of Hessian mercenaries because they were incapacitated after an alcohol-fueled revel. Yet he continued to insist that the benefits of hard alcohol consumption for military organizations were universally recognized and not to be disputed, advising Congress to establish public distilleries to ensure a reliable supply of rum to the fledgling U.S. Army.[49] Despite this strange commitment to liquid poison, the United States, and the United States Army, has ended up doing rather well.

Equally surprising is the central role that the production and consumption of intoxicants play in cultural life, from ancient to modern times. All over the world, wherever you find people, you find ridiculous amounts of time, wealth, and effort dedicated to the sole purpose of getting high. In ancient Sumer, it is estimated that the production of beer, a cornerstone of ritual and everyday life, sucked up almost *half* of overall grain production.[50] A significant portion of the Incan Empire's organized labor was directed toward the production and distribution of the corn-based intoxicant *chicha*.[51] Even ancient dead people were obsessed with getting wasted. It is hard to find a culture that did not send off their dead with copious quantities of alcohol, cannabis, or other intoxicants. Chinese tombs from the Shang Dynasty were packed with elaborate wine vessels of every shape and size, in both pottery and bronze.[52] This represented a cultural investment equivalent, in today's terms, to burying a few brand-new Mercedes SUVs in the ground with their trunks full of vintage Burgundy. Ancient Egyptian elites, the world's first wine snobs, were sent off in tombs full of jars that carefully recorded the vintage, quality, and name of their content's maker.[53] Because of its centrality in human life, economic and political power has often been grounded in the ability to produce or supply intoxicants. The Incan emperor's monopoly on *chicha* production both symbolized and reinforced his political dominance. In early colonial Australia, power was so thoroughly dependent on the control and distribution of rum that the first building in New South Wales was "a secure booze-bunker," guarding the precious imported liquid that also served as the New South Wales's primary currency.[54]

This pairing of civilization and fermentation has therefore been a constant theme in human history. Our earliest myths equate drinking with becoming properly human. As we have seen, Sumerian myth portrays the joy created by beer as key to transforming the animalistic Enkidu into a human. In ancient Egyptian mythology, the supreme god Ra, angered by something people have done, orders the fierce lion-headed goddess Hathor to completely destroy humanity. After she happily begins her rampage, Ra takes pity on humans and decides to call her off, but Hathor won't listen. Ra only manages to get her to stand down by tricking her into drinking a lake of beer, dyed red to resemble human blood. She gets wasted and falls asleep. "And thus," observes Mark Forsyth, "mankind was saved by beer."[55]

The expansion of cultures can also be tracked by following the waft of alcohol. Commenting on the settling of the American frontier, Mark Twain famously characterized whiskey as the "earliest pioneer of civilization," ahead of the railway, newspaper, and missionary.[56] By far the most technologically advanced and valuable artifacts found in early European settlements in the New World were copper stills, imported at great cost and worth more than their weight in gold.[57] As the writer Michael Pollan has argued, Johnny Appleseed, whom American mythology now portrays as intent on spreading the gift of wholesome, vitamin-filled apples to hungry settlers, was in fact "the American Dionysus," bringing badly needed *alcohol* to the frontier. Johnny's apples, so desperately sought out by American homesteaders, were not meant to be eaten at the table, but rather used to make cider and "applejack" liquor.[58]

The cultural centrality of intoxication remains to this day. A traditional household in the South American Andes, for instance, is still dominated by the various pots needed to produce *chicha* from corn, a process that requires multiple days and produces a beverage that then spoils quickly.[59] (So much for the preservation theory... [60]) A significant portion of an Andean woman's workday is dedicated solely to keeping up the supply; the same is true of millet beer in Africa, the production of which defines gender roles and dominates agricultural and household rhythms.[61] In kava cultures in Oceania, the production of this intoxicating tuber monopolizes huge swaths of both arable land and agricultural labor, and its consumption dominates social and ritual events.[62] When it comes to market economies, contemporary households around the world officially report spending on alcohol and cigarettes at least a third of what they spend on *food;* in some countries (Ireland, Czech Republic) this rises to a half or more.[63] Given the prevalence of black markets and underreporting on the topic, actual expenditures must be significantly higher. This should astound us. It's a lot of money to be spending on an evolutionary mistake.

Moreover, as mistakes go, this one is personally and socially costly as well as expensive. In Oceania, kava consumption leads to widespread negative health consequences, ranging from hangovers to dermatitis to serious liver damage. Alcohol is worse. For the year 2014, a Canadian research institute estimated that the annual economic cost of alcohol consumption, including impacts on health, law enforcement, and economic productivity, was $14.6 *billion*—quite a lot for a country of Canada's size. This includes 14,800 deaths, 87,900 hospital admissions, and 139,000 years of productive life lost.[64] The American Centers for Disease Control (CDC) estimates that, from 2006 to 2010, excessive drinking led to 8,000 deaths annually, 2.5 million years of potential life lost, and $249 billion in economic damage. In 2018, a widely publicized article in the British medical journal *Lancet* concluded that alcohol use ranks among the most serious risk factors for human health worldwide, playing a role in almost 10 percent of global deaths among fifteen- to forty-nine-year-olds. "The widely held view of the health benefits of alcohol needs revising," it concluded, "particularly as improved methods and analyses continue to show how much alcohol use contributes to global death and disability. *Our results show that the safest level of drinking is none.*"[65]

Given the dangers of intoxicant consumption, we should sympathize with the anguish and confusion of Shakespeare's Cassio, fired for drunkenness by an angry Othello after getting tricked into overindulgence by the sneaky Iago:

> O thou invisible spirit of wine,
> if thou hast no name to be known by,
> let us call thee devil!...
> O God, that men should put an enemy
> in their mouths to steal away their brains!
> that we should, with joy, pleasance, revel
> and applause, transform ourselves into beasts![66]

Why do we voluntarily poison our minds? That we continue to so actively and enthusiastically transform ourselves into beasts, despite all of the terrible costs involved, is a puzzle that is even more surprising in light of the type of creatures we actually are. Those other subverters of our brains, pornography and junk food, enjoy free rein because humans have, as yet, no ready defenses against them. The case of intoxicants is different. Unlike other species, humans have both genetic and cultural defenses against this enemy in the mouth that steals one's brain. It is worth considering these in some detail.

A GENETIC MYSTERY: WE ARE APES BUILT TO GET HIGH

Plenty of animals get drunk by accident. From fruit flies to birds, monkeys to bats, many animals are attracted to alcohol, and often

41

to their profound detriment.[67] For instance, family lore has it that an illegal pet lemur owned by my relatives in Bologna, Italy, became addicted to the rubbing alcohol that one of them, a midwife, had around the house in large quantities. One day the unfortunate creature broke into a bag of alcohol-soaked cotton swabs, became wildly drunk, and fell to his death from the top-floor apartment balcony. There are similar stories of inebriated birds breaking their necks flying into windows or simply taking naps on cat-patrolled lawns. Perhaps most dramatic are stories of drunken elephants running amok, trampling and destroying everything in their inebriated path.

It is also not unheard of for human beings to suffer the same fate as the Bolognese lemur. The number of drunken *Homo sapiens* who have fallen to their deaths is certainly non-zero. Yet it is important to see that we are not constrained, the way other animals are, to occasional raids on the alcohol swabs. Indeed, the very existence of concentrated doses of alcohol owes its existence to us.[68] And yet, as far as I know, Bolognese midwives have never been tempted to get wasted on rubbing alcohol. They, like everyone else around them, were constantly surrounded by effectively unlimited amounts of alcohol, in multiple forms of varying degrees of tastiness. Given such easy access, it should surprise us how few intoxicated people fall to their deaths from Bologna apartment-block balconies. The tastiness and potency of the local red wines alone would cause us to expect a constant stream of corpses to be piling up in apartment courtyards throughout the province, not to mention the excellent grappa. But, to my knowledge, this unfortunate lemur is the only recorded Bolognese alcohol-related falling death, at least in that particular apartment complex. Imagine a world inhabited by billions of lemurs or elephants with opposable thumbs, huge brains, technology, and an endless supply of high-potency alcoholic beverages: It would be chaos and carnage on a scale one shudders to imagine. And yet this is not the world we live in.

This is partly because our particular lineage of ape appears to be genetically adapted to processing alcohol and eliminating it quickly from the body. Alcohol dehydrogenases (ADH), which are produced by many animals, especially those that feed heavily on fruit, are a class of enzymes involved in the processing of ethanol, the alcohol molecule. A small set of primates, including humans, possess a super-powered variant of ADH, called ADH4. In the animals that possess it, this enzyme is the first line of defense against alcohol, quickly breaking ethanol down into chemicals that can be readily used or eliminated by the body. One theory holds that this enzyme variant gave a crucial evolutionary edge to the African ancestor of modern apes (gorillas, chimpanzees, and humans). This ancestral ape, possibly in response to competition from monkeys, had moved from living in trees to foraging on the ground. ADH4 allowed it to make use of a new, valuable food source: fallen, over-ripe fruit.[69] This calls into question any overly simplistic version of the hijack theory of intoxicant use.

The evolutionary anthropologist Ed Hagen and colleagues[70] have similarly shown that when it comes to plant-based recreational drugs like cannabis or hallucinogens, the hijack theory, at least, is undermined by evidence that humans have biologically adapted to consuming them. Take cannabis, for example. THC, the ingredient in cannabis that gets you high, is actually a bitter neurotoxin produced by the plant to avoid getting eaten. All plant drugs, including caffeine, nicotine, and cocaine, are bitter for a reason. The astringent taste is a message to herbivores: Back off, if you eat this it's going to hurt your stomach or mess with your brain and probably both. Most herbivores, being sensible, give plants like this a wide berth. However, some particularly stubborn ones—or those with a powerful taste for coke—develop countermeasures, evolving to produce enzymes that detoxify the toxicants. It is significant that humans appear to have inherited these ancient mammalian defenses to plant toxins, suggesting that plant-based drugs, like alcohol, are not an evolutionarily novel scourge, but rather a longtime friend.[71]

Another way to put this is that we are animals built to get high. This fact makes hijack theories less plausible, suggesting that alcohol and other intoxicants have long been part of the adaptive environment within which we evolved, rather than a recent and unanticipated threat. This still leaves hangover theories on the table, though. We may be biologically preadapted to handle the relatively low alcohol levels found in rotting fruit, or to processing the toxins found in the coca leaf, but this leaves us helpless once the development of agriculture, large-scale societies, technology, and trade puts powerful beers, wines, and distilled spirits at our disposal, or tempts us with refined cocaine or super-THC strains of cannabis. The ancient Scythians, fearsome warriors though they were, would have been reduced to dribbling idiots had they had access to the Maui Wowie or Bubba Kush I can pick up at my local cannabis dispensary. Hangover theories allow for ancient adaptations to intoxicants, ones we share with other species, but assume that the unique changes experienced by *Homo sapiens* over the last 9,000 years or so—catapulting us from small-scale hunter-gatherer life to that of globalized urbanites—happened too fast for genetic evolution to catch up.

This is not a safe assumption. It is commonly thought that genetic evolution takes a long time to work, producing adaptations only over time scales on the order of hundreds of thousands or millions of years. Given that humans have only been living in large-scale societies for something like 8,000 to 10,000 years, this would mean that humans have remained genetically unchanged since we were hunter-gathers roaming the Pleistocene African plains. Another common belief is that, since the advent of large-scale societies and the invention of agriculture, humans have cast off the shackles of day-to-day survival challenges, and thereby freed themselves from the pressures of genetic evolution.

Neither of these beliefs is true. For instance, people from cultures that raise cattle have, sometime in the last 8,000 years, genetically adapted to digesting milk as adults. The Tibetan plateau, with an average elevation of 4,500 meters, is an incredibly harsh environment. But sometime between 12,000 and 8,000 years ago, its inhabitants began developing genetic adaptations that protect against the harmful effects of the low oxygen levels found there. Similarly, fisherfolk in Southeast Asia who rely on diving in the ocean to obtain food have, over the past couple thousand years, evolved the ability to hold their breath for long periods of time.[72] So there has been plenty of time since the advent of agriculture for adaptations against the misuse of alcohol to evolve. If hangover theories of human intoxicant use were true, we would expect that genetic evolution would be working overtime to eliminate our taste for getting high. We would also expect that any human population that evolved a defense against this "enemy in the mouth" would be very successful, causing the relevant genes to spread rapidly to any region of the world where high-powered intoxicants are available.

Of course, genetic evolution can sometimes be pretty stupid, as work-arounds like masturbation and junk food attest. There are also many problems that genetic evolution simply cannot help us with. Consider the human spine. It is a terrible design for an upright, bipedal organism, which is why so many people suffer from lower back problems. Yet evolution does not have the luxury of designing us from scratch. It has to do the best it can with what it has been given, a body scheme designed for climbing and living in trees, gradually modified and hacked until it could walk upright.[73] Natural selection cannot peer around corners or see beyond adaptive valleys, and is often stuck in the ruts of evolutionary pathways that were originally chosen for long-irrelevant reasons. It is theoretically possible, then, that our taste for alcohol is like our achy lower backs, an unfortunate example of how genetic evolution is so constrained by previous decisions that it effectively has its hands tied. Evolutionary biologists call this "path dependence." It is also the case that selection cannot act on a mutation that doesn't exist. So, another possibility is that a cure for our taste for intoxication is biologically possible, but the spinning of the genetic mutation roulette wheel has yet to land upon it. This would be a simple availability problem.

At least when it comes to our taste for alcohol, both path dependence and availability problems can be definitively ruled out. This is because an excellent solution to this supposed evolutionary mistake, this parasite of the human mind, already exists in the human gene pool, and has for a really long time.

We've mentioned the enzyme ADH, the body's first line of defense against poisons like alcohol. ADH takes the ethanol molecule, C_2H_6O, and strips off a couple of the hydrogen atoms—hence its name, "alcohol dehydrogenase." The resulting molecule, C_2H_4O or acetaldehyde, is still quite toxic, and definitely not something you want floating around in your body. This is where a second liver enzyme, aldehyde dehydrogenase (ALDH), takes over. Through a process of oxidation (adding an oxygen atom pulled from a passing water molecule), it converts acetaldehyde into acetic acid, a much less dangerous chemical, and one that can in turn be easily changed into water and carbon dioxide and eliminated from the body (Figure 1.2).

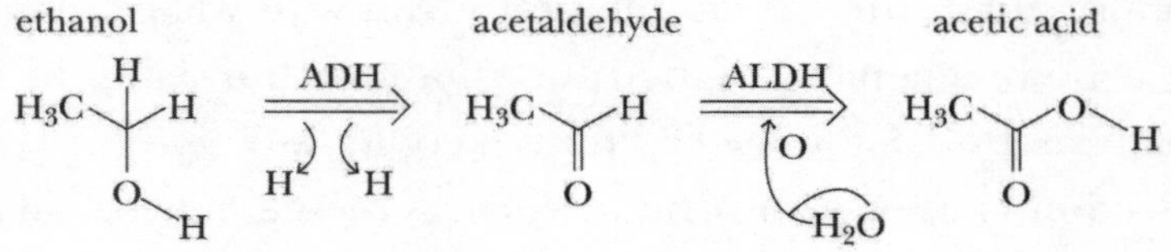

Figure 1.2. Conversion of ethanol to acetaldehyde and then acetic acid by ADH and ALDH.

Things get ugly when this second step is delayed. If ADH is happily converting alcohol into acetaldehyde, but ALDH is slacking on its job, acetaldehyde starts to build up in the body. This is bad. The body signals its displeasure and alarm by producing facial flushing, hives, nausea, heart palpitations, and difficulty breathing. The message to us is: Whatever it is you are doing, stop right now. The worst-case scenario would be one where ADH is really good at its job, producing lots and lots of acetaldehyde, but ALDH is unusually bad at its job, letting this toxic substance build up and start spilling over everywhere, like a hapless Charlie Chaplin on the assembly line. Surprisingly, considering that the genes coding for these two enzymes are not directly linked, this odd coupling of super-efficient ADH and terrible-slacker ALDH does appear in some human populations. It is most common in East Asians, which is why the condition it causes is sometimes known as the "Asian flushing syndrome." It also seems to have evolved independently in parts of the Middle East and Europe.

The body is not stupid. The symptoms produced by excessive acetaldehyde are so unpleasant that people who experience them listen up and quickly learn to avoid putting large quantities of alcohol into their bodies. In fact, the flushing reaction makes alcohol consumption so aversive that a drug able to induce it in genetically typical individuals is used to treat alcoholism.[74] Carriers of the gene that codes for these variant enzymes, as well as functionally analogous mutations found in pockets of non-Asian populations, have been effectively freed from their desire for alcohol. They might drink in moderation, and so get to enjoy whatever benefits moderate drinking affords—anti-microbial treatment, trace minerals and vitamins, and calories, if those are in short supply. However, the dramatic suite of unpleasant physical symptoms that ensues when they drink too much for their inefficient ALDH enzymes to handle means that they are protected from the excesses of drunkenness and alcoholism. They get to eat their cake without any danger of ending up facedown in it. What an awesome solution to the hijacking or hangover problem! It is as if there were a gene that made pornography unappealing while leaving one's drive for reproductive sex unimpaired, or made Twinkies taste like cardboard but broccoli like the most delectable ambrosia. Quite a genetic coup.

This genetic silver bullet for the alcohol problem has been knocking around in the human gene pool for a long time, as far back as 7,000 to 10,000 years ago in East Asia. Interestingly, its distribution in East Asia seems to track the appearance and spread of rice-based agriculture. This may indicate a response to the sudden availability of rice wine,[75] but some theories posit that its original adaptive function was to protect against fungal poisoning.[76] Hunter-gatherers eat wild vegetables, fruit, and meat, and don't go in much for storing food. Once you have rice farming, though, you have large quantities of grains that, when socked away for later in a damp environment, quickly get invaded by fungi. High concentrations of acetaldehyde in the body, while unpleasant to experience, are also very effective in killing off fungal infections. So it might be the case that the flushing reaction paid for itself, as it were, by allowing moderate drinkers who experienced it to also safely consume stored rice. Others observe that the inefficient form of ALDH seems to guard against tuberculosis, and might have been selected for because of the enhanced risk of disease once people began living in the large, dense groups made possible by agriculture.[77] In either case—fungicide or anti-TB medicine—the protective effect of heightened acetaldehyde against alcoholism would be merely a nice side effect.

But what a great side effect! If alcohol consumption were merely a counterproductive accident of our evolutionary history, we'd expect the "Asian flushing" genes to spread like wildfire anywhere

where excessive alcohol consumption was a potential problem. In other words, almost everywhere in the civilized world. Given the rapidity with which other novel genetic adaptations, such as lactose tolerance or performance at high elevations, have taken over in regions where they are useful, anyone able to read this book should flush after a drink or two.[78]

This is clearly not the case. The genes producing this reaction remain confined to a relatively small region of East Asia, and are not even universal there. The versions that evolved independently in the Middle East and Europe have remained similarly constrained in scope. When genetic evolution solves a serious problem, it is not shy about sharing. The fact that the miracle "cure" for our taste for getting high seems to have relatively few takers calls any evolutionary mistake theory into serious question.

A CULTURAL MYSTERY: PROHIBITION'S STRANGE FAILURE TO TAKE OVER THE WORLD

In the year 921, an Islamic scholar named Ahmad Ibn Fadlan was sent by the caliph of Baghdad on a diplomatic/religious mission to the Volga Bulgars. These were recent converts to Islam, living on the banks of the Volga River in what is now Russia, and apparently the caliph felt that their grasp of their new faith could use a little tune-up.

Along the way the embassy encountered a group of Vikings, who impressed Ibn Fadlan with their height and physiques, but horrified him with their disgusting personal habits, orgiastic funeral ceremonies, and out-of-control drinking. "They drink the mead to insensibility, day and night," he writes, "It often happens that one of them dies with the cup still in his hand."[79]

49

The Vikings were seriously into alcohol. The name of their chief god, Odin, means "the ecstatic one" or "the drunken one," and he was said to subsist on nothing but wine. Mark Forsyth points out the significance of this: While many cultures have a god of alcohol or drunkenness, in order to give alcohol some recognized role within society, when it comes to the Vikings the chief god and the god of alcohol are one and the same. "That's because alcohol and drunkenness didn't need to find their place *within* Viking society, they were Viking society. Alcohol was authority, alcohol was family, alcohol was wisdom, alcohol was poetry, alcohol was military service, and alcohol was fate."[80]

This had its downsides as a cultural strategy. Medieval Vikings would make modern frat boys look like herbal tea-sipping grannies. As Iain Gately notes, binge drinking played such a central role in their culture that "a striking number of their heroes and kings died from alcohol-related accidents,"[81] ranging from drowning in enormous vats of ale to being slaughtered by rivals while rolling about in a drunken stupor. Perpetually drunken and heavily armed warriors also posed a threat to those around them. The highest praise accorded to the legendary Viking/Anglo-Saxon hero Beowulf was that "he never killed his friends when he was drunk." As Forsyth observes, "This was clearly something of an achievement— a thing so extraordinary that you'd mention it in a poem."[82] On top of these more dramatic and violent downsides, Viking society also had to endure the enormous material costs of intoxicant production and long-term health consequences of heavy drinking, such as cancer and liver damage.

The incredible costs of alcohol, in terms of material expense, health consequences, and social disorder, have been very much in the forefront of the minds of anti-drinking campaigners of all stripes throughout history. Prohibitionist literature goes at least as far back as second millennium BCE China. A poem from the *Book of Odes,* called "When the Guests First Take Their Seats," gives voice to a lament familiar to anyone who has hosted a dinner party that has gone on a bit too long:

> When the guests first take their seats,
> How mild and decorous they are!....
> Those who are drunk behave badly;
> Those who are not feel ashamed.

A later ode warns the notoriously hard-drinking last kings of the Shang Dynasty, "Heaven did not let you indulge in wine / And follow ways against virtue."[83] Traditional Chinese historiographers argue that it was precisely excessive drinking and womanizing that led to the downfall of the dynasty. Reflecting upon their behavior, a member of the Western Zhou Dynasty (1046 to 771 BCE), which replaced the Shang, was inspired (supposedly) to give a speech entitled, "Against Drinking Wine," in which he complained about their alcoholism, sexual vice, and neglect of ritual duties. In place of the smell of fragrant and proper sacrifices to the ancestors, in the last years of the Shang nothing ascended to Heaven other than "the people's grievances and the rank alcoholic odor of drunken officials."[84] Heaven was not pleased, and called in the Zhou people to carry out the Shang's destruction.

China has been worried about alcohol ever since.[85] In their myths, they attributed prohibitionist policies to their earliest sage-kings. The legendary Yu, supposed founder of the Xia Dynasty (traditional dates 2205 to 1766 BCE), is said to have sampled some wine, relished its taste, and then promptly exiled the woman who made it for him. Wine should be banned, he is reported to have said, because it "would one day destroy someone's kingdom."[86] China is also responsible for what are probably the earliest recorded attempts to legally impose prohibition as public policy.

The "Against Drinking Wine" speech goes a step further than exile, declaring that anyone caught drinking wine should be put to death. The origins of this document are unclear, but we have evidence of similar proclamations from bronze items definitely datable to the early Zhou period,[87] and later Chinese rulers issued a steady stream of political edicts against drinking.[88]

Ancient Greece combined an appreciation for the social usefulness of moderate drinking with a contempt for drunkards and strong warnings about the dangers of alcoholic excess. One early playwright puts advice concerning the virtues of moderation and sobriety into the mouth of the god of wine, Dionysus, himself:

> Three cups only do I propose for sensible men, one for health, the second for love and pleasure and the third for sleep; when this has been drunk up, wise guests make for home. The fourth cup is mine no longer, but belongs to hubris; the fifth to shouting; the sixth to revel; the seventh to black eyes; the eighth to summonses; the ninth to bile; and the tenth to madness and people tossing the furniture about.[89]

Later, in the West, various forms of Christianity waged a long war against drinking, sometimes under the blanket term of "gluttony," one of the seven cardinal sins. Today we tend to think of gluttony in terms of overeating, and the sin certainly does cover having one pork chop too many. But excessive drinking was not only traditionally covered by moralistic anti-vice diatribes, it was often their primary focus. "The list of the possible effects of the sin of gluttony," notes one scholar of fifteenth-century penitence manuals, "included talkativeness, unseemly joy, loss of reason, gambling, unchaste thoughts, and evil words." These vices, she wryly notes, "would not seem to follow from overeating."[90] The more recent anti-drinking crusader William Booth, the founder of the Salvation

Army, declared that "the drink difficulty lies at the root of every-thing. Nine-tenths of our poverty, squalor, vice, and crime springs from this poisonous tap-root. Many of our social evils, which over-shadow the land like so many upas trees, would dwindle away and die if they were not constantly watered with strong drink."[91] Today we pass our lives in blissful ignorance of the danger presented by upas trees, native to Southeast Asia and supposedly so poisonous that their mere smell could kill, but the message is clear. Drinking is bad.

Given the evident costs of intoxication, it is not surprising that many political leaders have seen complete abstinence as the secret to cultural success. For instance, the early twentieth century Czech thinker, independence leader, and first president of Czecho-slovakia, Tomáš Masaryk, saw abstinence as key to the liberation of the Czech people. In a statement directed at his notoriously hard-drinking compatriots, he declared that "a nation which drinks more will undoubtedly succumb to one that is more sober. The future of each nation and especially of a small nation depends on...whether it stops drinking."[92]

Anyone who has ever been to that part of the world can attest that the Czechs didn't stop drinking. In fact, they continue to hold the honor of drinking more beer per capita than any other nation-ality, and consistently rank among the highest in per capita overall alcohol consumption in the world.[93] And yet the Czech Republic, despite its brief subjugation to the equally hard-drinking U.S.S.R., has yet to be wiped off the map. Prohibition also never got off the ground in China—the same Zhou Dynasty tombs containing bronze tripods declaring death to anyone who consumes alcohol are also chock-full of elaborate and expensive wine vessels, and there was never a successful attempt to limit alcohol consumption. Yet Chinese culture has had a pretty long run of it. The booze-sodden Vikings, dismissed by the abstinent Ibn Fadlan as dirty drunkards, were also wildly successful as a cultural group. They dominated and terrified huge swaths of Europe, discovered and colonized Iceland and Greenland, became the first Europeans to reach the New World, and ended up siring a good proportion of modern Northern Europeans. A loose attitude toward alcohol consumption doesn't seem to slow cultural groups down very much.

This is even more puzzling than the Asian flushing gene's failure to sweep through the world. As Tomáš Masaryk saw clearly, a culture that spends entire evenings consuming liquid neurotoxins—created at great expense and to the detriment of nutritious food production—should be at an enormous disadvantage compared to cultural groups that eschew intoxicants altogether.

Such groups exist, and have for quite some time. Perhaps the most salient example is the Islamic world, which produced Ibn Fadlan. Prohibition was not a feature of the earliest period of Islam, but according to one *hadith*, or tradition, it was the consequence of a particular dinner at which companions of Mohammed became too inebriated to properly say their prayers. In any case, by the end of the Prophetic era in 632 CE, a complete ban on alcohol was settled Islamic law. It cannot be denied that, in the cultural evolution game, Islam has been extremely successful. From its origins among nomadic tribes on the Arabian Peninsula, it has become one of the great world religions, dominating vast swaths of the Eurasian continent and South and Southeast Asia. Nonetheless, Islam continues to have to rub shoulders with alcohol-friendly faiths such as Christianity and Confucianism (not to mention the Vikings), when both the hijack and hangover theories would attribute to it a decisive advantage in the cultural evolutionary game.

Even more damaging for any non-adaptive theory of intoxicant use, the situation on the ground with regard to Islam is much more complicated than theology would have it. First of all, the ban on *khamr*, or intoxicants, is often interpreted to apply only to alcoholic beverages, or even only to alcohol fermented from grapes or dates, leaving other intoxicants untouched. Most prominent among these alternate intoxicants is cannabis, usually in the form of hashish. This was beloved in particular by the somewhat heretical Sufis, but also widely tolerated in the general population.[94] Moreover, despite theological prohibition, Islamic cultures have historically varied considerably in how strictly they enforce the ban on alcohol. In most Islamic cultures, alcohol consumption has been allowed in the private home, especially among elites, and in some places and times it has even played a prominent role in public life. As one historian observes, "Throughout history, Muslim rulers and their courtiers have consumed alcohol, often in huge quantities and sometimes in public view; the examples of ordinary Muslims violating their religion's ban on drinking are too numerous to count...The Islamic proscription of alcohol was a gradual, almost reluctant process, one that reveals itself as relative despite its apparent absoluteness, providing loopholes, allowing for subterfuge, and leaving open the chance of having one's guilt absolved."[95] It is worth noting that the Islamic world has given us our word for "alcohol" (from Arabic al-kohl) and first accounts of alcoholic distillation, as well as some of our greatest wine poetry. The celebrated Hafez of Shiraz, writing in the fourteenth century, went so far as to declare wine drinking to be the very essence of being human: "Wine has flowed in my veins like blood / Learn to be dissolute; be kind—this is far better than / To be a beast that won't drink wine and can't become a man."[96] If a ban on alcohol were a cultural evolutionary killer app, you'd expect it to be more consistently enforced.

Another teetotaling culture worth mentioning is the Church of the Latter-Day Saints, more colloquially known as the Mormons. Like Mohammed, Joseph Smith, the founder of Mormonism, came a bit late to the prohibitionist game. The *Book of Mormon* shares the generic Christian view of wine as a sacramental substance, and portrays at least mild intoxication as a genuine pleasure approved by God. The early Mormon church liberally employed wine at religious gatherings, even combining alcohol-driven feasting and dancing in the temple itself. It was not until Joseph Smith's 1833 revelation, called the "Word of Wisdom," that Mormons were told that God didn't want them consuming alcohol, caffeinated beverages, or tobacco. Alcohol use was then suppressed, but only gradually; total abstinence did not become official church doctrine until 1951.[97] It is fair to say, however, that the modern Mormon Church has taken up prohibition with impressive zeal.

The Mormons, then, seem like a group that is serious about eliminating mind hijacking chemicals from our lives, which should in turn give it a massive advantage over other groups. And the Mormon faith is, in fact, quite a success story. Although in recent years its rate of growth has slowed somewhat, it continues to out-pace global population increase, which is more than can be said for most religious faiths.

The very zeal and comprehensiveness of the Mormon church's war against psychoactives, however, should give us a clue about its actual function. The Mormon combination of the ban on Coca-Cola and coffee with its alcohol prohibition makes little sense if its main target is the cost of intoxication. Unlike alcohol and other intoxicating drugs, caffeine would seem to have only positive bene-fits for both individual faith and group success. Legend is that tea drinking arose among otherwise teetotaler Buddhist monks in Asia in order to help them maintain long periods of meditation, and without coffee and nicotine it is hard to say how many members of Alcoholics Anonymous would be able to make it through a meeting. Indeed, modern life would arguably grind to a sudden halt without cigarettes, coffee, and tea.

As the historian of American religion Robert Fuller has argued, the Mormon ban on psychoactive chemicals seems less targeted at the specific problem of alcohol, and more "a strategy to emphasize difference from other existing religious groups."[98] Similar arguments have been advanced about Islamic abstinence, which may have originally functioned to distinguish the early Muslim world from the wine-drinking cultures of the Mediterranean and Near East that surrounded it.[99] Prohibition is a dramatic cultural statement, serving as a powerful group marker and costly loyalty-inspiring display. In the case of the Mormon Church, this ability to distinguish themselves from others through abstinence has been combined with other creative and impressive practices, such as demanding a two-year mission from all male devotees and allowing the proxy baptism of long-dead ancestors. It is likely this package of cultural evolutionary innovations, rather than the ban on alcohol itself, that accounts for the relative success of the Mormon faith.

To summarize, if intoxication had overall negative effects on cultural groups, we would expect anti-intoxicant norms to become universal, especially since cultural evolution moves much faster than genetic evolution. If alcohol bans are in the process of taking over the world, however, they are certainly taking their time about it. How do we explain the failure of prohibition in ancient China or the United States, or the continued existence of, say, France? Groups that have officially banned chemical intoxicants often wink at private use or look the other way when elites indulge in public. Many who are more serious about banning intoxication, such as Pentecostals or Sufis, replace the joys of drunkenness with some form of non-chemical ecstasy, such as speaking in tongues or ecstatic dance. This all suggests that intoxication is performing a crucial functional role in society. This would make it resistant to being eliminated by cultural fiat, and would create a vacuum that needs to be filled in the rare cases where it is genuinely taken out of the picture.

PICKLES FOR THE ANCESTORS?

Our earliest written records from ancient China, the so-called "oracle bones" dating back to the Shang Dynasty, provide us with insight into early Chinese ritual-religious life. *Jiu* (酒, "wine")—a broad term referring to a millet-derived ale, but possibly also drinks involving wild grapes and other fruits—features prominently, holding pride of place in sacred ritual sacrifices. Indeed, the religious historian Poo Mu-chou observes that, although various foodstuffs were also burned and sacrificed to the gods and ancestors, wine was so central that its use was synonymous with the ceremony itself, and the character for ritual offering (*dian*, 奠) seems to portray a wine jar placed on a stand.[100] One poem from the *Book of Songs,* perhaps the most ancient of our transmitted documents from China, purports to describe an ancient Zhou ritual held in celebration of an abundant harvest:

We make wine and sweet liquor
As offerings to the ancestral spirits of earth and grains
Together with other sacrificial items,
To bring down blessings broadly to all.[101]

The focus of the ceremony is the "wine and sweet liquor," which seem particularly beloved of the ancestral spirits. There are also other items being sacrificed, presumably various foodstuffs, but it's hard to know: All we are told about is the booze, and then, you know, yada yada yada, some other stuff. This is typical of early China, where ritual celebrations and offerings to the spirits focus exclusively on the consumption and offering of alcoholic beverages.[102]

Chinese culture is no outlier in this regard. Throughout history and across the world, alcohol and other intoxicants—kava, cannabis, magic mushrooms, hallucinogen-laced tobacco—tend to be the prime offering in sacrifices to the ancestors and gods, as well as the central focus of both everyday and formal communal rituals. The most dramatic artifacts in the graves of iron-age elites in Europe were enormous drinking vessels,[103] and Egyptian ancestors demanded sacrifices of wine from their descendants. At Seder dinner, a cup of wine is left for Elijah; he'd presumably be disappointed to arrive and find only a dry piece of matzo at his seat. As Griffith Edwards, author of *Alcohol: The World's Favorite Drug*, notes, social toasts are always made with alcoholic beverages, and seem to derive some of their power from their intoxicating essence. "With 'To your health!' we have the most everyday and pervasive example of a drinking ritual with a whiff of magic." He further observes that "the necessity of alcohol for this ritual is a widespread and ancient assumption," quoting the Victorian journalist and author Edward Spencer Mott: "Do we express our unfeigned joy and thankfulness for having a great and good Queen to reign over us by toasting her in flat soda water? Forbid the deed!"[104]

This all should mystify us more than it does. Banquets and religious rituals centered on kimchee and yogurt would provide all of the proposed benefits of alcohol with none of the costs. Spirits should be perfectly happy with some nice, nutritious pickles instead of a poisonous, bitter beverage. Yet no culture on the planet offers pickles to the ancestors, and the world has yet to see the rise of a teetotaling, kimchee-based super civilization. This strongly suggests that there is something special about alcohol, and more to the function of intoxication, than we have realized.

What might this function be? We cannot answer this question without understanding the problems for which intoxication represents a solution. Humans are the only animal that deliberately and methodically gets high. We are also very unusual in a variety of other ways. As we'll see in the following chapter, those of us who live in agriculture-based civilizations are even stranger. In order to unravel the evolutionary mystery of our taste for intoxication, we need to get a sense of the unique challenges that confront humans—selfish apes who appear to behave, at least on the surface, like selfless social insects.

CHAPTER TWO

LEAVING THE DOOR OPEN
FOR DIONYSUS

A chimpanzee allowed to compete on the reality show *Survivor* might be expected to destroy its competition. Not just literally—a grown chimp has massive teeth and is strong enough to tear a person apart—but also in terms of survival skills. Chimps are smart, tough, and extremely clever problem-solvers. If I had to put my money on the likely survivor of a massive, multi-species parachute drop of individuals into an unfamiliar wilderness, I'd bet on a chimp. A human wouldn't even make the final five. Lone humans tossed into novel environments have a pretty short half-life.[1] Nevertheless, a chimpanzee contestant on *Survivor* would, in fact, be one of the first to get tossed off the island, at least after "the merge," when the two competing tribes are combined into one.

That's because human beings have a steep advantage: We spend most of our time in a large-scale version of the merge, where surviving is not primarily about strength or individual cleverness, but rather about social skills. It certainly doesn't hurt if you are strong or good at building fires or catching game, but the people who ultimately emerge at the top of the *Survivor* heap tend to be coalition builders, alliance negotiators, and judicious manipulators.[2] For a very long time, the primary adaptive challenge for human beings has been other human beings, not the physical environment. Knowing how to find water in the desert is important, but nowhere near as crucial as learning how to share that water with other humans, negotiating the division of labor in carting the water back to camp, and sussing out who is likely to try to steal your share of the water when you're not looking.

This observation is crucial to unraveling the puzzle of why we get drunk. Humans are the only species that deliberately, systematically, and regularly gets drunk. The rarity of this behavior is not surprising, given its costs. What is surprising is why humans nonetheless persist in doing it. As we've seen, our taste for intoxication doesn't seem to be an evolutionary accident, given its persistence in the face of counterpressures and the existence of both genetic and cultural "solutions." Both hijack and hangover theories seem inadequate as explanations. But this still leaves open the question of what intoxication is good for.

To answer this question, we first need to understand the specific ways in which it's hard being human. Species arise and survive by adapting to a particular *ecological niche*. This term refers partly to a species' place in the local ecosystem, whether it is predator or prey, herbivore or carnivore. More fundamentally, it refers to its repertoire of methods for succeeding at occupying its place, securing food and shelter, hiding or hunting, and dealing with both fellow species members and other species. The gradual changes wrought in populations as they adapt to a novel niche is one of the processes by which new species arise. As niche environments drive specialization, things can get pretty weird.[3]

Consider the Mexican tetra, a small freshwater fish popular among aquarium owners. This species has diverged into two dramatically different forms, as certain sub-populations have come to live exclusively in underground caves rather than surface rivers. Cave tetras have gradually adapted to this lightless environment by becoming pale white and, more dramatically, by losing their eyes.

Pigment is useful in sunlit waters, where it helps fish blend into their visual backgrounds. In the same way, eyes and the neural machinery required to run them more than pay for themselves in the surface world, where they are essential for locating prey and identifying predators. In the dark world of the cave, however, pigment and vision are of no help, so adaptive pressure favored individuals who dispensed with these physiologically expensive, but now useless, features. The blind, pale cave tetra, although somewhat bizarre-looking, is exquisitely adapted to its new, dark ecological niche, where it efficiently pursues prey by smell and touch. There is no going back now, however: Thrown into the light- and color-filled world of a surface river, the cave tetra would be instant dim sum. Having adapted to the cave, it has to stay there.

Among primates, humans are in a situation not unlike the cave tetra's. *Homo sapiens* have achieved their impressive success by adapting to an extreme and unusual ecological niche, one very different from that inhabited by our primate ancestors and closest primate relatives today. In the same way that the cave tetra can no longer survive out in the bright, terrifying world of the surface river, humans have become so dependent upon culture that we can no longer live without it.[4]

For instance, one of the earliest, and most basic, technologies to which we've adapted as a species is fire. As the primatologist Richard Wrangham observes, fire is useful in a lot of ways, not least of which is that it allows us to cook vegetables and meat.[5] Cooked food is easier to eat and digest, which means that the first humans or proto-humans to master fire no longer needed the massive jaws, robust teeth, and elaborate digestive system that chimpanzees, for instance, need to handle their diet of rough, fibrous fruits and raw meat. This allowed early humans to redirect physiological resources to beefing up other parts of their anatomy, such as the energy-hungry brain. Like the eyeless cave tetra, this loss makes us more efficient in our new environment of cooked, and therefore pre-digested, food, but it also makes us dependent upon fire. In adapting to the ecological niche that included fire use, our line of hominids impaired its ability to survive on raw food alone. (Contemporary raw foodists have yet to get the memo on this.)

So, one feature of the "cave" to which humans have adapted is that it provides fire, among other basic cultural technologies. It also provides language and incredibly valuable cultural information, which explains the multiple human adaptations to mastering languages and learning from others. Compared with the environment to which our line of primates originally adapted, our cave is crowded and full of strangers, non-relatives with whom we need to somehow cooperate. Living there is cognitively demanding, requiring not only the ability to master a slew of artificial cultural technologies and norms, but also a capacity for producing novel ones.

Living in this niche therefore requires both individual and collective creativity, intensive cooperation, a tolerance for strangers and crowds, and a degree of openness and trust that is entirely unmatched among our closest primate relatives. Compared to fiercely individualistic and relentlessly competitive chimpanzees, for instance, we are like goofy, tail-wagging puppies. We are almost painfully docile, desperately in need of affection and social contact, and wildly vulnerable to exploitation. As Sarah Blaffer Hrdy, an anthropologist and primatologist, notes, it is remarkable that hundreds of people will cram themselves shoulder to shoulder into a tiny airplane, obediently fasten their seat belts, eat their packets of stale crackers, watch movies and read magazines and chat politely with their neighbors, and then file peacefully off at the other end. If you packed a similar number of chimpanzees onto a plane, what you'd end up with at the other end is a long metal tube full of blood and dismembered body parts.[6] Humans are powerful in groups precisely because we are weak as individuals, pathetically eager to connect with one another, and utterly dependent on the group for survival.

I have compared humans to eyeless cave tetras and puppies, but in this respect a different analogy is more apropos: that of social insects, such as ants or bees.[7] Compared to other primates, we are freakishly social and cooperative; not only do we sit obediently on airplanes, we labor collectively to build houses, specialize in different skills, and live lives that are driven by our specific role in the group.

This is quite a trick for a primate to pull off, considering our most recent evolutionary history. Hive life is (literally) a no-brainer for ants: They share the same genes, so sacrificing for the common good is not really a sacrifice—if I'm an ant, the common good simply *is* my good. Humans, though, are apes, evolved to cooperate only in a limited way with close relatives and perhaps fellow tribe members, acutely alert to the dangers of being manipulated, misled, or exploited by others. And yet we march in parades, sit in obedient rows reciting lessons, conform to social norms, and sometimes sacrifice our lives for the common good with an enthusiasm that would put a soldier ant to shame. Trying to hammer a square primate peg into a circular social insect hole is bound to be difficult. But, as we'll see, intoxication can help.

THE HUMAN ECOLOGICAL NICHE:
CREATIVE, CULTURAL, COMMUNAL

Chickens are not as stupid as you might think. Descended from red jungle fowl, a species native to Southeast Asia, they have suffered surprisingly few ill effects, cognitively speaking, from the process of domestication. Your average farmed chicken is more or less as smart as her wild cousin, capable of working with simple numbers and logical relationships, reasoning about cause and effect, taking the perspective of others, and experiencing empathy.[8]

All of these impressive cognitive abilities and behaviors, however, are innate. Chickens aren't stupid, but they are rigid and dull—what they are able to do at two weeks old is pretty much all they will ever be able to do. This is not surprising given the fact that chickens are a type of bird categorized by biologists as "precocial." They pop out of the egg fully formed, feathered, and ready to go, their little heads already packed with everything they will need to know about the relatively narrow ecological niche to which they are adapted. This means that they can hit the ground running, as it were, which has obvious benefits.

Other bird species, so-called "altricial" birds, are more or less helpless at birth. They emerge from their eggs naked and blind, unable to move on their own or feed themselves, and with the prospect of flying only a distant dream. They are completely unable to survive without intense parental investment, often for relatively long periods of time. New Caledonian crows, for instance, require two full years of care before they can fully fend for themselves, and they often hang around their parents for up to four years, mooching food and picking up skills. Given that crows generally only live into their teens, this represents a surprisingly large chunk of their lifespan.[9]

At first glance, the chicken strategy seems much better. Why, as a species, saddle yourself with helpless young and clingy teenagers who continue to steal milk from the fridge and leave their dirty laundry about? Given the obvious advantages of popping out of the egg ready to go out into the world, it is difficult to see how or why the altricial strategy evolved, or why species that started out altricial haven't all evolved into precocial ones.

There are downsides, however, to peaking too early, as many

high-school homecoming kings and queens have discovered to their detriment. Undersized, picked-on Dungeons & Dragons–playing geeks often end up as highly educated, well-traveled, and successful adults. Similarly, weak, naked, newly hatched crows—who would get shoved into lockers and have their lunch money stolen by their scary chicken peers—eventually turn into startlingly creative animals with enormous behavioral flexibility.

Crows, for instance, are members of a class of bird known as "corvids," which also includes ravens and jays. Corvids are capable of making tools that require several, cumulative steps to create (such as carefully shaped hooks or leaves cut into particular shapes), carrying these tools with them on foraging expeditions (evidence of foresight and planning), and using them to dislodge insects from hard-to-reach places.[10] They have impressive memories, evinced by their ability to hide or "cache" excess food over a broad geographical range. Perhaps most surprisingly, they display impressive social intelligence. A corvid that is observed by another corvid while caching food will often wait until the potential thief is distracted and then go back to re-hide its bounty. They will also, when being observed by another corvid, hide pretend food, like small stones that look like nuts, or lead would-be spies on wild-goose chases away from the real location of their cached food. (They ignore chickens, for obvious reasons.) A corvid would do pretty well on post-merge *Survivor*.

Crucially, corvids are flexible and creative, modifying these complex behaviors in response to novel conditions. In the laboratory, corvids deprived of their normal tool-making materials can manufacture hooks out of novel materials, such as metal wire. Put in conditions where a perishable food item, crickets, decayed more rapidly than in the wild, they quickly learned to cache and recover more durable peanuts instead. Like monkeys and apes, corvids are able to extract general rules from particular learning tasks, and apply these rules to analogous but novel situations. For example, when given a food reward after pecking a blue square to match a blue stimulus, they quickly learn the general rule, "match the stimulus," and can continue to follow it when the colors themselves are changed or even replaced with shapes.[11]

Corvids can also solve completely novel problems that require insight and imagination. In one laboratory experiment,[12] for instance, ravens were presented with a piece of meat attached to a string hanging from a perch. The only way to get at the meat was to pull the string up a bit with the beak, lay it down on the perch, hold the accumulated string down with a claw, and then carefully repeat the process six or eight times. Amazingly, one wild raven in the experiment, after carefully sizing up the situation, solved this task on the first try. The other ravens in the experiment figured it out after only a few attempts.

Faced with some food hanging out of reach on a piece of string, a hapless chicken would starve to death. In general, precocial species like chickens or pigeons are never able to exhibit anything other than a relatively narrow range of behavior. In the laboratory, they can learn specific tasks by rote but are unable to discern the more general rule behind them. This leaves them utterly flummoxed by novel problems. Having been trained to peck a blue square when presented with a blue square, a pigeon has no idea what to do when the colors are changed or are replaced with shapes. It cannot formulate the abstract concept of "matching." The heyday of precocial species was their youth, when they strutted around the playground without fear, too cool and successful to bother with books or school. This may not seem like a great long-term strategy, but it really depends on context. Both strategies—peaking early or peaking late—exist in the world because each has its own advantages, and you can't say which is likely to be better without knowing the environment in which it is to be implemented.

As the developmental psychologist Alison Gopnik and her colleagues have observed, general intelligence, behavioral flexibility, ability to solve novel problems, and a reliance on learning from others tends to roughly correlate with an extended period of helpless immaturity.[13] This relationship is found across a broad range of animals, including birds and mammals, suggesting that it tracks a fundamental evolutionary trade-off between narrow competence and creative flexibility. In other words, species as a whole seem to be placing their bets on either the late-blooming high-school geek or precocious homecoming queen strategy, and then moving into ecological niches where the strategy they have chosen provides the best payoff. Or, having found themselves thrown into an environment requiring one strategy or another, they specialize in it.

It should not surprise us that humans are outliers here, as in many domains. We are the uber-geeks, the picked-on dorks, the teacher's pets of the animal world. As any parent or grandparent is well aware, we are far and away the most helpless of altricial mammals. Our offspring are completely useless, and would get stomped on—metaphorically and literally—by their chimpanzee or monkey peers. Anyone who has ever waited impatiently at the front door while a four-year-old attempts to tie her shoes can be forgiven for wishing human children were more like chickens. It's not just the utter lack of dexterity or inability to remember the relevant steps that is so aggravating. Young humans are *spacey*—they get halfway through tying their shoes and then forget what it is they are supposed to be doing, focusing instead on working a booger out of their nose or untying the one shoe they'd already managed to get on. You look away for a moment to check the time and then glance back to see that not only are the shoes not on, but for no conceivable reason they've now decided (*ta-da!*) to take their *pants* off.

It's possible that the haplessness of our offspring explains another unusual fact about humans. We are one of the few species where females go through menopause—basically giving up the whole reproductive game—with many years still left to live. This is an odd thing for an organism to do, unless it's the case that it can maximize its overall reproductive success by forgoing personal reproduction, and instead plow its time and resources into helping with its grandchildren and great-grandchildren. This, in turn, only makes sense if these little ones are such an enormous hassle to deal with that they *need* grandmothers to survive. This appears to be the case with humans.[14] It takes a village to raise our uniquely weak, distractible, and aggravating offspring.

Humans have adopted such an extreme form of the peak-late strategy because, as a species, we have come to inhabit an equally extreme ecological niche. The main demands imposed upon us by the odd, crowded cave to which we have adapted can be summed up with what I'll call the Three Cs: we are required to be *creative, cultural,* and *communal.* The demands of the Three Cs make us, like the helpless, blind, altricial crow chicks, more vulnerable than robust and less complicated animals. For instance: sharks. You'd never want to put a four-year-old human up against a four-year-old shark. Yet it remains the fact that our weak, mewling infants grow into relative masters of the universe, putting sharks in aquariums, eating their fins in soups, and now, unfortunately, driving them to extinction in many regions of the world.

The human transition from extreme vulnerability to immense power is, however, a journey that is fraught with challenges. Understanding the nature of these challenges is crucial to understanding the potential adaptive advantages of intoxication. We get drunk because we are a weird species, the awkward losers of the animal world, and need all of the help we can get. Let us now look at the Three Cs and why extended childhood, or its chemical equivalent, might be extremely useful for a species like us.

70

THE CREATIVE ANIMAL

Oedipus really has trouble catching a break. The protagonist of Sophocles's tragedy *Oedipus Rex* is abandoned to die as an infant, forced out of his home city of Corinth because of an oracle that says he is fated to kill his father and marry his mother. Remarkably, he survives, but on his way to find someplace new to hang his hat, he is provoked into a road-rage-induced fight with an aggressive old man at a crossroads, killing the man and his attendants—and thus unknowingly fulfilling the first part of the prophecy. Worse is in store. Trying to enter Thebes, he is set upon by the horrible Sphinx, who is terrorizing the city and threatens to kill Oedipus and the citizens of Thebes if he cannot answer a riddle: "What walks upon four legs in the morning, two in the afternoon, and three in the evening?"

The answer, of course, is a human, crawling as a baby, then walking upright, and finally requiring the aid of a cane. Later, after he has become king of Thebes (and, we must note, also married his mother), Oedipus faces another crisis in the form of a terrible plague. Some, like the soothsayer Teresias, turn to the gods for guidance, hoping to discern the proper way forward from clues found in the flight of birds or other omens. Oedipus castigates them, recalling his encounter with the Sphinx:

> Tell us, has your mystic mummery ever approached the truth?
> When that hellcat the Sphinx was performing here,
> Tell us: What help were you to these people?
> Her magic was not for the first man who came along:
> It demanded a real exorcist. Your birds—
> What good were they? or the gods, for the matter of that?
> But I came by,
> Oedipus, the simple man, who knows nothing—
> I thought it out for myself, no birds helped me![15]

What vanquished the Sphinx was not magic or divine intervention, but the power of human creative insight.

As the cultural historian Johan Huizinga notes, riddles that must be solved, upon the pain of death, are a common feature across the world's sacred mythical cultures. "In its mythological or ritual context," he observes, "it is nearly always what German philologists know as the *Halsrätsel* or 'capital riddle,' which you either solve or forfeit your head. The player's life is at stake."[16] The universality of high-stakes riddles in human mythology highlights, in symbolic form, one of the main challenges that confronts us in adapting to our ecological niche: Humans need to be creative to survive.

As a species, we are uniquely dependent on the insights and inventions that give rise to cultural technologies, ranging from kayaks and harpoons to fish traps and longhouses.[17] We sew clothes, fashion multi-part tools, build shelters, process and cook our food. Most other species are simply given by nature what they need to get by: the lion its claws, the gazelle its speed. Hive-building insects and dam-building beavers are also on autopilot. Their artifacts might look superficially like human inventions, but they are really just an extension of their genome, no different from a bird's wing or shark's teeth. Even the raven who makes a hook out of a piece of wire is more or less following a script—inaccessible worm, need a hook—although it is flexibly able to make this tool out of whatever material is available. Humans genuinely invent *new* things, in the sense that cultural innovations are not simply read off our DNA. Faced with the problem of inaccessible worms, a truly human-like crow wouldn't just mess around with hooks, it would invent worm *farms* that would allow it to just reach in and grab what it wants. Humans transform the world through our creative technologies, and we cannot survive without them. The utter dependence of humans on creative insight is the real lesson of Oedipus's encounter with the Sphinx.

In reflecting on the riddle of the Sphinx, Huizinga, who passed away in 1945, did not have the benefit of modern cognitive science, but he understood the psychological challenge well enough. "The answer to an enigmatic question is not found by reflection or logical reasoning," he argued. "It comes quite literally as a sudden *solution*—a loosening of the tie by which the questioner holds you bound."[18] No amount of algorithmic chain reasoning or brute force can give you the solution to a riddle: You just need to relax your mind and see the answer in a flash of insight. Psychologists refer to this process, one that is aimed at producing an *aha* moment, as lateral thinking. Another task that requires lateral thinking is the Remote Associates Test (RAT). You are given three seemingly unrelated words, such as *fox, man,* and *peep,* and asked to think of a fourth word that unites them all. (See the endnotes for the answer.[19]) The Unusual Uses Test (UUT) similarly requires thinking outside the box: Given a common artifact, like a paperclip, the participant is asked, within a time limit, to come up with as many novel uses for it as possible (toothpick, earring, fishhook).

Lateral thinking tasks are actually quite fun, like solving riddles, and can be adapted into enjoyable party games. But as in the Oedipus myth, there is deadly seriousness behind our ability to solve them. Humans are like mature corvids with useless beaks and no wings. A crow uses tools occasionally, when it needs to get a particularly hard-to-reach grub or to access a deeply hidden bit of food. Even in the most low-tech societies, however, humans are completely helpless without tools and the creative insights that generate them. We need creativity simply to function.

Human beings' long developmental period, our extended childhood, may be one response to this need. If you want help with the Unusual Uses Test, just rope in a little kid. Solving lateral thinking tasks is where the four-year-olds who get distracted by an ant crawling across the floor when they are supposed to be putting on their shoes, or who suddenly decide to strip off their pants apropos of nothing, really come into their own. Kids are crappy at logistics and planning, but their little chaotic minds explore the nooks and crannies of possibility space with a speed and unpredictability that leave adults completely in the dust. Look at any small child, at any point in the day, and they are probably performing something like the UUT: turning a cardboard tube into a rocket ship, or using a large stick as a horse.

In fact, one of Gopnik's most important arguments is that this cognitive flexibility and creativity is a *design feature* of youth. She and her colleagues review evidence that suggests that when it comes to novel learning tasks, the young of many species often outperform their elders.[20] This is certainly true of humans. One of Gopnik's experiments involved introducing subjects to a "blicket detector," a roughly shoebox-sized device that lights up and plays music when exposed to "blicketness." They were asked to place a variety of different-shaped objects on the device to figure out which of them possessed this elusive quality. The default assumption among adults was that blicketness should be a property of a single object, and they did almost as well as young children in the "disjunctive" condition where this was the case. In the more counterintuitive "conjunctive" condition, the box would only light up when a particular combination of objects was placed on it at the same time— in these trials, one had to see that "blicket" did not, as one would expect, refer to any single object in isolation. In this condition, four-year-olds completely crush their elders. Roughly 90 percent of them successfully identify the conjunctive blicket, as opposed to around 30 percent of adults, and performance declines gradually with age (Figure 2.1).

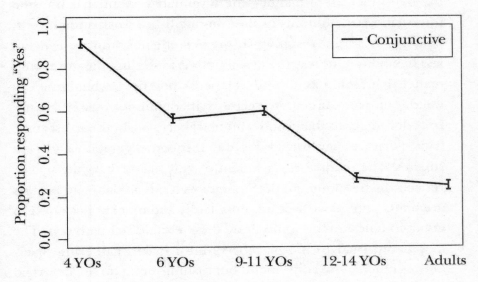

Figure 2.1. Proportion of subjects, by age, correctly identifying a blicket in the conjunctive condition, where a blicket consists of two separate objects.[21]

What explains this difference in performance, as well as its change over time? Take a look at this decline in performance compared to another trend, this one from developmental neuroscience, showing the relative densities of gray and white matter in the human frontal cortex (Figure 2.2).

While we might imagine that the brain matures through accumulation, building up more and more neurons in a given region, maturation in fact results from what is called "neural pruning," the gradual elimination of unnecessary neural connections. A region of the brain becomes mature when it settles down into a lean, functionally well-organized system. A good proxy for neural pruning in the brain is the relative density of gray versus white matter in a given region. Gray matter, the neuron-rich part of the brain that does the bulk of the computational work, decreases in density as a region matures. As gray matter density decreases, the density of white matter—the myelinated axons that transmit information, the outputs of the computational work done by gray matter—increases, resulting in greater efficiency and speed but less flexibility. One way to envision this is to see an immature, gray-matter-rich region as an undeveloped, open field, where one can wander in many directions unconstrained, but not very efficiently. In order to get to that wonderful blackberry bush to harvest some fruit, I have to bushwhack my way through vegetation and ford streams. The gradual replacement of gray matter by white matter reflects the development of this field: As roads are laid and bridges are built, I can move around more easily and quickly, but now I'm going to tend to move only along these established pathways. The new paved road to the blackberry bush makes gathering black-berries much more convenient, but rushing along on the new road I will miss the delicious wild strawberries I would have otherwise stumbled upon in the brush. There is a trade-off between flexibility and efficiency, between discovery and goal achievement.

As the brain develops, gray matter density decreases and white matter increases in a linear fashion, reflecting increasing maturity and functional efficiency. The relevant portion of this region for our purposes is the prefrontal cortex (PFC), the seat of both abstract reasoning and what psychologists refer to as "cognitive control," the ability to remain focused on task, resist distractions and temptations, and regulate emotions. As we can see from Figure 2.2, it takes a long time for the frontal cortex, including the PFC, to finish its process of neural pruning. It is, in fact, the PFC that is the last region of the brain to mature, not reaching its adult state until the early twenties. This is why the teen years are so dangerous: Teens have adult-like motivational systems, raging hormones, and access to dangerous technologies like automobiles, but only limited rational self-control.

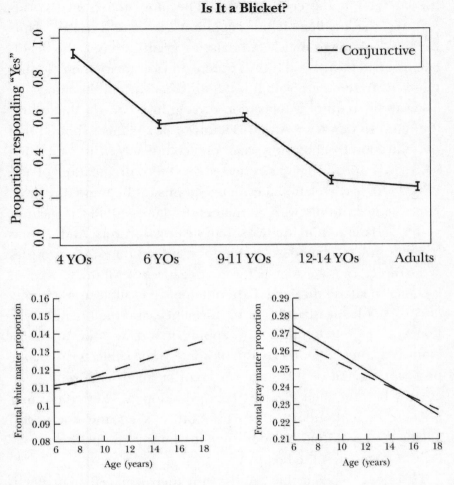

Figure 2.2. Successful performance in counterintuitive conjunctive condition by age (top), compared to increase in white matter density (bottom left) and decline in gray matter density (bottom right) in the frontal cortex over the course of development (dashed lines represent male subjects and solid lines females).[22]

The way in which the trends graphed in Figure 2.2 correlate is significant. As people age, we see gray matter density decreasing and white matter density increasing in the frontal cortex, and this reflects a corresponding decline in performance on the lateral-thinking creativity task. The more the frontal cortex matures, the less flexible our cognition becomes. The PFC, while key for remaining on task and delaying gratification, is the deadly enemy of creativity. It allows us to remain laser-focused on task but blinds us to remote possibilities. Both creativity and learning new associations require a relaxation of cognitive control that allows the mind to wander.[23] An fMRI study of jazz pianists showed that the transition from playing scales or a completely written-out tune to freely improvising was reflected by a downregulation of the PFC.[24] Other correlational evidence points in the same direction. For instance, adults with permanently damaged PFCs perform better on lateral-thinking tasks than healthy controls. And, thanks to the wonders of modern technology, at least one study provides some direct, causal evidence for the negative role of the PFC when it comes to lateral thinking. Experimenters had subjects perform a creativity task, measured their performance, and then temporarily knocked their PFC offline by zapping it with a powerful trans-cranial magnet.[25] (Don't try this at home.) The subjects did better post-zapping. All of this data suggests that small children are so creative because their PFCs are barely developed. There is nothing policing their thoughts, which has both upsides and downsides. Taking twenty minutes to put on your shoes is the price you pay for thinking out of the box.

This doesn't mean that adults, with their lean, efficient PFCs, are completely useless when it comes to creativity and innovation. Like jazz pianists easing into a stretch of improvisation, adults are able to sometimes relax the vigilance of their PFC and surrender themselves to play. In this regard, full-grown humans are still childlike—or, at least, potentially childlike. Huizinga, the cultural historian so intrigued by riddles, famously argued that what is distinctive of humans is our desire to play. In this sense, we resemble domesticated dogs.

One of the reasons that dogs strike us as so cute is that, relative to their wolf ancestors, they display "neoteny," the extension of juvenile characteristics into adulthood. In other words, even as adults they look and act like wolf pups, with rounded, puppy-like features, an intense desire to play, and a readiness to trust. As the play researcher Stuart Brown notes, adult human beings, in both our appearance and playfulness, are essentially the "Labradors of the primate world."[26] We look, and behave, more like chimp babies than chimp adults. Species that display neoteny (like dogs) tend to be more flexible but less efficient and self-sufficient; those showing mature characteristics (like wolves) are brutally efficient but rigid. In the same way that an extended period of childhood tends to predict cognitive flexibility, across a broad range of species there appears to be a positive correlation between brain size and playfulness.[27] Human children, like dog pups, are thus doubly immature: youthful versions of a species that itself retains youthful characteristics.

Even as adults we still enjoy playing—maybe not as much as our children would like, but way more than wolves or adult chimps do. This helps with the creativity challenge. As Brown observes, many important inventions—steam engines, airplanes, clocks, firearms— started as toys.[28] Playing around, coming up with things to amuse our children and ourselves, can help us to regain childlike creativity. Even *imagining* ourselves as children seems to help. One study found that undergraduates did better on a creativity test when first asked to think about how their seven-year-old selves would respond to a canceled school day. Reminiscing about impromptu fort building or idly throwing rocks into a stream seems to free up our ability to think laterally.[29] Play is also crucial to learning, which brings us to our second C: culture.

THE CULTURAL ANIMAL

Human beings' individual creativity, while impressive in itself, is enormously magnified and crucially enhanced by our ability to accumulate and build upon the insights of the past through the medium of transmitted culture, or cultural innovations that are preserved and passed down. This is obvious in modern, high-tech cultures. The iPhone that I carry around in my pocket represents hundreds of years of accumulated R&D, ranging from its basic operating principles to the materials from which it is made. Crucially, no single human could hope to produce even the most basic component of an iPhone, or any complex cultural technology, through sheer insight or creativity. Innovations are always necessarily gradual and incremental, building on the accumulated insights of past humans. We are the cultural animal par excellence, and our ability to share the products of our individual creativity and pass them on to future generations is the key to our ecological dominance.[30]

Moreover, cultures as a whole can figure out the solutions to problems that are, in principle, beyond the capacity of any single individual to solve. As cultural evolutionary theorist Michael Muthukrishna and colleagues argue, we need to think of our brains not just as individual organs sitting in our heads, but as part of an extended network, nodes in a massive "collective brain."[31] Creative breakthroughs often arise in this network through a process larger and more powerful than any given individual could reproduce. "Innovations, large or small," they write, "do not require heroic geniuses any more than your thoughts hinge on a particular neuron. Rather, just as thoughts are an emergent property of neurons firing in our neural networks, innovations arise as an emergent consequence of our species' psychology applied within our societies and social networks. Our societies and social networks act as collective brains."[32]

To take one, relatively low-tech example of the collective brain solving a problem beyond the capacity of an individual brain to figure out, consider the case of manioc (or cassava). As the anthropologist Joseph Henrich explains,[33] this tuber is an important staple crop that was first domesticated in the Americas, but it cannot be simply cooked and eaten like a potato. Most varieties contain a bitter substance, a defense against insects and herbivores, that causes cyanide poisoning when the plant is consumed. Cultures that have historically relied upon manioc have therefore developed elaborate, multiday procedures to process the root, involving scraping, grating, soaking, and boiling it, and then patiently waiting for several days before baking and eating it. Modern chemical analysis shows that this process dramatically reduces manioc's toxicity. *Why* unprocessed manioc is dangerous, and how processing and leaching render it safe, is something we've only come to understand quite recently. Yet ancient cultures solved the problem of unlocking the food potential of manioc millennia ago, through a long, blind process of trial and error combined with cultural memory. Groups that did something right, initially by accident—for instance, letting the manioc soak for a few days because they forgot about it—did better than groups that didn't make this fortunate mistake. Other groups began to emulate these more successful groups. Over time, the accumulation of useful errors or random variations gave rise to the set of culinary practices that allowed manioc to be safely consumed.

It is important to realize that no single individual could have figured this out; the long time lag between consuming manioc and experiencing its negative effects, as well as the challenge of assembling the various necessary detoxification steps in their proper order, make it exceedingly unlikely, if not entirely impossible, that anyone could have learned to detoxify manioc alone. Moreover, in cases like this, the people benefiting from the cultural solution generally have no idea how or why it works, or even that it is necessary in the first place. As Henrich notes, the procedure of manioc detoxification is *causally opaque* to any given individual. If you think you need to soak the manioc for two days because your mom told you that the ancestors will get angry if you don't, you've misunderstood the causality, but who cares—the relevant fact is that you rendered the manioc edible.

As Henrich also observes, a particular historical experiment displays the danger of trying to wing it in the absence of traditional cultural memory. In the early seventeenth century, the Portuguese, noting that manioc is easy to grow and provides impressive yields even in marginal cropland, imported it to Africa from South America. It quickly spread to become an important staple crop in the region, and still is today. Yet the Portuguese neglected to also import the South American indigenous *cultural knowledge* about how to properly detoxify manioc. The difficulty of reinventing the wheel, as it were, is dramatically illustrated by the fact that many contemporary Africans, hundreds of years after the introduction of manioc, continue to suffer from health problems caused by low-level cyanide poisoning.[34] The point, Henrich concludes, is that "cultural evolution is often much smarter than we are."

An anthropological survey of island cultures across the Pacific showed that population size and connectedness with other islands correlated positively with the number of tools possessed by a culture, as well as its degree of tool complexity. In modern urban societies, increased population density leads to increased innovation, as measured by proxies such as number of new patents or R&D activity per capita.[35] Cultural accumulation allows not only the gradual building-up of technology and knowledge, but also creates a virtuous circle where existing cultural resources become raw material for new, individual inventions. With the invention of agriculture and advent of large-scale civilizations, this virtuous circle went into hyperdrive. Sharing across massive empires united scores of local ethnicities and ecosystems, all trading raw materials, cultural knowledge, and technology with one another. This process of cultural evolution has given us automobiles, airplanes, high-speed elevators, and the internet.

This human reliance on culture is very unusual in the animal world. Most species deal with the world by means of "asocial learning," whereby a single individual sizes up a problem and formulates a solution to it. Our closest biological relative, the chimpanzee, relies almost entirely on asocial learning. Humans, however, at some point crossed a kind of evolutionary Rubicon.[36] The increasing advantages of accumulated culture began to reshape our brains to become more and more dependent on "social learning," a process in which individuals, when confronted by a problem, draw upon solutions provided by their culture. In order to take advantage of this information, individuals need to be open and trusting, willing to rely on others rather than go it alone.

Muthukrishna and his colleagues ran iterations of a computer model of the real world where they varied starting conditions for a host of biological and environmental parameters, including brain size, group size, length of juvenile period, mating structure, richness of the environment, etc., and observed which learning strategies emerged as the dominant ones. As Figure 2.3 illustrates, under most conditions, selection favors asocial learners, with some models producing a mild reliance on social learning. It is only under a narrow range of conditions, where groups increase in size, brains become large, juvenile periods grow long, and cumulative cultural knowledge expands and provides adaptive advantages, that we see a spike in models producing individuals who rely almost exclusively on social learning.

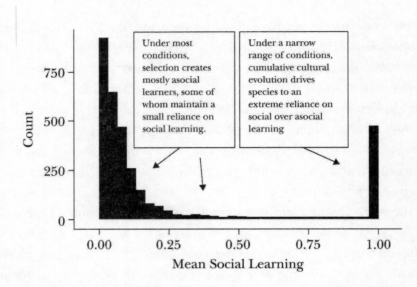

Figure 2.3. Degree of mean social learning that emerged as the dominant strategy in various models; number of models represented on y-axis.[37]

Note the huge valley between asocial and social learning that humans came to traverse, as well as the narrow corner of adaptive space into which we have been pushed. Once social learning becomes valuable enough, a species that has access to it will be driven inexorably away from asocial learning and become completely dependent on culture.

The power of cumulative cultural evolution has reshaped us as dramatically as the blind cave tetra. Once we bet all of our chips on cultural learning, there was no going back to asocial, individual learning. There is a common image of human innovators or pioneers as isolated, bold individuals, wresting solutions to the puzzles presented by nature through sheer willpower and insight. This ideal of the lone genius may well describe an innovative chimpanzee or crow, but is nonsense when it comes to humans. Chimpanzees are strong and independent and smart; humans are weak, dependent on others, and, as individuals, no rocket scientists. Like the cave tetra, we are elaborately adapted to life in the dark, sheltered cave of social learning, but would be blind and helpless if thrown into a world without culture.

Our dependence on culture means that our minds need to be open to others, so that we can learn from them. This is another realm where extended childhood is clearly an adaptation to our ecological niche. Babies and children are the most powerful learning machines on the planet. As Alison Gopnik observes, "The evolutionary imperative for babies is to learn as much as they can as quickly as possible."[38] This means that their under-developed PFC is, as she puts it, a functional design feature rather than flaw. Babies and young children are easily distracted, but also aware of a much broader spectrum of what is happening around them, paying attention to incidental details that would escape the attention of focused, goal-oriented adults.[39] Moreover, children's incessant playing and messing around with things has the incidental effect of allowing them to learn skills and develop a sense of the causal structure of the world around them.[40] The causality they need to understand is not just physical but also social. For instance, my daughter's favorite game in her early toddlerhood was tea party. She would sit me down in a circle, along with a menagerie of stuffed animals, and force me to wear a tiara. (Not my best look.) As the host, she would pour us all (pretend) tea, offer us (pretend) snacks, and engage in (nonsensical) polite conversation. She and her classmates, when left to their own devices, would similarly "play" all sorts of social scenarios—teachers and students, doctors and patients, parents and children.

All this playacting is not just for fun, although children are designed to have fun doing it. It is a serious and crucial process of learning the causal structure of the social world around them. A drive to play, and an openness to soaking up information from the people around them, is a feature of childhood designed to allow children to acquire the accumulated culture they need to survive. It is mind-boggling how much information human children need to master: their local language(s), as well as with whom to speak which language; how to dress, eat, cook, hunt, build, paddle, track game; local social structures and norms, taboos, rituals, myths.

As anyone who has tried to learn a foreign language after the age of thirteen or so can easily attest, this ability to learn atrophies as we enter adulthood. It is not just languages that we have trouble picking up as we mature. Adults have trouble learning novel social practices and norms—and indeed they're typically resistant to doing so. Presented with Chinese food in Iowa, most locals demand a fork instead of chopsticks and expect a sweetness level that approximates their current diet. Having reached maturity outside of a British cultural context, I will always find Marmite disgusting. Grown-ups are similarly rubbish at acquiring new skills. I learned to play tennis as an adult, and even after years of lessons and playing have trouble hitting a forehand with proper form. My daughter learned to swing as a child with effortless perfection, and will soon be running me ragged on the court.

Here again, as with the creativity decline, we can blame the PFC. There is a body of evidence showing that already-acquired complex, skilled behavior is run by implicit, automatic systems, and that bringing the PFC and executive control online really screws things up. The best way to sabotage a professional tennis player's serve is to ask them to think about how they are doing it as they do it. Asking a group of people engaged in effortless, pleasurable banter to reflect on their social dynamics is guaranteed to ruin the party. This is why having a fully developed PFC makes you relatively impervious to new knowledge and skills. And this is why the PFC takes so long to mature and childhood extends so long in humans: We have an enormous laundry list of things to learn from the people around us, so we need to remain flexible and receptive for as long as possible.

Having been shaped by evolution to become unusually open to and dependent upon learning from others, humans also needed to learn how to play well with others, as befits the Labradors of the primate world. Compared to other primates, we *are* like goofy dogs: bizarrely tolerant of strangers, open to new experiences, ready to play. This openness to others, while necessary for our success, also creates vulnerability. We *need* others to a degree that is not true of any other primate. This leads us to the third C, our intensely communal nature.

THE COMMUNAL ANIMAL

Life arises from cooperation. The biological world presents us with a dazzling kaleidoscope of intertwined cooperating units, from the level of genes all the way up to cells and organisms and social groups. The chromosomes in your body represent what can be seen as a "society of genes,"[41] a collection of individual segments of DNA that are dependent upon one another and share a common fate. The cells built by these chromosomes "agree," metaphorically speaking, to specialize into different tissue and organ types, all with an eye toward working together to make the individual organism to whom they have entrusted their fate more successful in getting at least half of its package of genes into the next generation.[42]

Once we get up to the level of individual organisms, these cooperating units might go it alone, battling the world or other organisms, or choose to team up with other cooperative units. Sometimes, in the latter case, the degree of cooperation involved is so intense that groups of cooperating individuals begin to look like superorganisms, replicating on a social scale the same sort of cooperative agreements that make individual bodies possible.[43] In social insects like bees and ants, for instance, individuals quickly specialize into distinct functional castes, such as workers or soldiers or reproductive queens. A worker ant will mindlessly and selflessly toil away at acquiring food for others, while a soldier ant will readily and heedlessly sacrifice itself to neutralize intruders. All that matters is that the queen survives to ensure that the group's genes get into the next generation.

Primates are more selfish. They generally don't go in for self-sterilization or heedless suicide. As we have noted, though, humans—the social insects of the primate world—are an exception. The degree to which we depend on, and cooperate with, one another to achieve things completely beyond our individual abilities looks a bit like bees or ants, with their impressive hives and complicated divisions of labor. But our primate biology leaves us with an evolutionary problem—at a deep level, we nonetheless remain selfish, backstabbing apes. A queen bee never has to worry about insubordination on the part of her subjects. Human rulers get poisoned or decapitated or simply voted out of office all the time, as our set of personal desires, our chimpanzee DNA, rears its individualistic head.

The tension between our need to cooperate on a large scale and our individual, primate-driven selfishness appears clearly in the sort of dilemmas inherent to social cooperation. Any time tension arises between the public good and individual interests there is the danger of what economists call "defection," a situation in which a selfish individual benefits from the public good while not contributing to it. This tension goes by many names, including the "tragedy of the commons" or the "free-rider problem."[44] Fish populations diminishing? It would be better if everyone agreed to fish less, but on the open seas how do you enforce this? No one wants to be the sucker who stayed home while her rivals went out and scarfed up the last of the bluefin tuna. So the bluefin tuna gets driven to extinction. Enjoy using the communal kitchen at work to heat up or prepare your lunch? In the absence of a clearly defined and enforceable cleaning schedule, such spaces quickly decline into disgusting, unusable cesspools in a way that is untrue of privately controlled spaces. That is because it is in no one's individual interest to clean if others are not pulling their own weight—if I give in and wipe down the increasingly slimy sink or finally empty the dishwasher, I am allowing others to free ride on my effort.

These sorts of cooperation challenges pervade the human social world on every scale of interaction. They hamper global efforts to combat climate change, cause political parties and economic cartels to fracture,[45] and often force individuals into difficult choices. One such choice is the basis of the famous Prisoner's Dilemma thought experiment, which vividly illustrates a variant of the free-rider problem. Imagine you have been detained and accused of committing a crime. The prosecutor tells you that another suspect in the same crime, of whom you know nothing, has also been detained. You are offered a deal: Rat out the other person and you will get a slap on the wrist, a one-month sentence, while the other person will get the full three-year sentence. Refuse to talk, and you will be charged with obstruction of justice and given a six-month sentence. You know that, if you both end up accusing the other, you'll both be charged as accessories in the crime and sentenced to two years. You have no way of communicating with the other prisoner.

It is helpful to view this dilemma in terms of a payoff matrix (Table 2.1).

	B stays silent (co-operates)	B accuses A (defects)
A stays silent (co-operates)	6 months for both	3 years for A, 1 month for B
A accuses B (defects)	3 years for B, 1 month for A	2 years for both

Table 2.1. Prisoner's Dilemma payoff matrix.

It would be in the overall best interest of both prisoners to remain silent and take the reduced sentence. However, there is no way to guarantee that the other prisoner will cooperate. Given the danger of you cooperating and being betrayed by the other, the only rational strategy is to defect, or accuse the other. This result is a suboptimal outcome for both individuals, but is the only safe strategy to play. Purely rational, self-interested individuals cannot win at the Prisoner's Dilemma.

Fortunately, human beings are not rational, or at least not primarily rational.[46] Most cooperation theorists agree that people are often able to cooperate in the face of a Prisoner's Dilemma because they are *emotionally committed* to one another. When we are bound to another person or to a group emotionally, through love or loyalty or friendship, we can trust others, cut through the dilemma, and thereby achieve the optimal outcome for everyone. Faced by a DA tempting them to rat out their buddy, real-life gang members are able to keep their mouths shut and get away with reduced sentences. This is not merely out of fear of retribution (*snitches get stitches*), but more importantly about loyalty to the group and the internal sense of shame that would come with defection. Nobody likes a snitch, and nobody feels good about *being* a snitch.

Crucially, the only reason this works is that we do not have complete conscious control over our emotions. In the moment, it is in my narrow self-interest to make up an excuse for why I can't get up early Sunday morning to help my friend move her couch, but doing so would make me *feel* guilty, so I drag myself out of bed, hangover and all, and go get my van. A health crisis, career wobble, or the simple overwhelming trauma of raising children might make a relationship momentarily disadvantageous for one or both partners, but irrational love serves as glue keeping romantic couples bound to one another through thick and thin. Social emotions, when sincerely felt, allow us to override the selfishness of our short-term, calculating mind, but only because we can't consciously *control* them. If we could, our conscious, rational mind would simply shut them down when it was in our self-interest to do so, and they would lose their efficacy. Love or honor that I can switch on or off when convenient is not true love or honor.

Here, again, it is the prefrontal cortex—that seat of abstract thought, instrumental reason, and cognitive control—that is the enemy. Two prisoners under the sway of their PFCs are always going to get stuck with two years behind bars. The only way to get off with the reduced sentence is to overwhelm the PFC with an irrational emotion like honor or shame. A pair of Captain Kirks can both win at the Prisoner's Dilemma; a couple of Mr. Spocks are going to rot in jail.

In order to understand how emotions can help us resolve cooperation dilemmas, and why our relative inability to consciously control them in the moment is crucial to their social function, another story from Greek myth is useful. (We are leaning on

Greek mythology a lot in this chapter.) One of the many dangers that confront the hero Odysseus in his wide-ranging travels is a passage near the island of the sirens. All sensible sailors steer well clear of these dangerous creatures, who use their seductive song to lure vessels into the shoals and then feast upon the helpless, shipwrecked sailors. Odysseus, however, is never one to shrink from an adventure. A consummate hedonist, he is keen to hear the siren song, which is said to be unimaginably beautiful. He is also well aware of the dangers. In his typically clever fashion, he comes up with a work-around, a trick to prevent his future self from "defection," from getting itself into trouble. He instructs his sailors to plug up their ears with wax so they cannot hear the dangerous temptation at all, and then to firmly tie him to the mast. In this manner, Odysseus is thus able to hear the sirens while being physically restrained from leaping to his death—which, in the moment of hearing the song, he dearly wishes to do.

The ropes binding Odysseus to the mast are a literal instantiation of what we can call a "pre-commitment." As the Cornell economist Robert Frank puts it, social emotions represent "passions within reason."[47] Love, honor, shame, and righteous anger only seem irrational: It is actually in our long-term rational interest to be at the mercy of uncontrollable emotions in the moment, when the situation demands it. Like the ropes binding Odysseus, emotions are only able to function as commitment devices because we can't simply will ourselves free of them. By falling in love or pledging sincere loyalty to a group, we are effectively tying ourselves to the mast, binding ourselves into emotional commitments that will restrain us from betraying others when temptation inevitably calls. This is a remarkably effective strategy, and it explains why so many romantic couples remain committed to one another, college kids help each other move couches, and imprisoned gang members indignantly refuse to snitch to the DA.

The need for trust in human relationships is obvious when it comes to Prisoner's Dilemmas. It is also clear in other non-transactional relationships that are driven by commitment and interdependence—classically, parent-child relationships, and those between the sick and their caregivers.[48] What is less commonly realized is the degree to which even interactions that seem purely transactional on the surface can occur only against a deeper background of implicit trust. When I pay $4 for a hot dog from a street vendor, the money-for-wiener trade rests upon a set of assumptions so long it would be impossible to exhaustively list. The hot dog is properly cooked. It has not been deliberately contaminated. The dollar bills I am handing over are not counterfeit. The hot dog contains (at least mostly) beef or pork, not dog meat. None of this is explicitly spelled out, but it is all nonetheless firmly taken for granted. This is also why the occasional discovered violation of one of these items of background faith is so scandalous: *Dog meat in local vendor's hot dogs! Local father passing funny money at the park!* Lurid tabloid headlines merely reinforce how deeply we trust these fundamental background assumptions, and how rarely they are violated.

When it comes to trust and communal bonding, as with creativity and cultural learning, children wildly outperform their elders. Children come into the world with a raw, almost desperate need to bond with and trust fellow members of their cultural group. This is apparent to anyone who has been approached by a voluble four-year-old in a crowded airport waiting area and subjected to a lengthy, high-speed introduction to the doll they are carrying. It is, in fact, the eager willingness of children to trust and interact with others, including complete strangers, that makes violations of that trust so tragic and horrific. Similarly, the absence of trustingness in a child is a sign that something has gone badly wrong in their environment.

When it comes to trust, as with creativity and learning, play is important. Throughout the animal world, play allows practice for important adult skills, such as hunting or fighting, and gives young individuals an opportunity to work out social hierarchy structures. Crucially, however, it also provides practice in trust. As the animal behavior expert Marc Bekoff has observed, play typically involves deliberate vulnerability—think of a playful dog baring its belly or neck—as well as the signaling of trustworthiness. The "play-bow" with which dogs greet each other in public before engaging in a wrestling match is a social trust signal: If you bow too, we have agreed to enter a play world where our bites will not be deep, our growls will not be serious, and we will take turns dominating the other.[49] Although researchers have long thought that the primary function of play was for skill practice and training, this socializing and trust-building function seems more fundamental. As Stuart Brown observes, "Cats deprived of play-fighting can hunt just fine. What they can't do—what they never learn to do—is to socialize successfully. Cats and other social mammals such as rats will, if seriously deprived of opportunities for play, have an inability to clearly delineate friend from foe, miscue on social signaling, and either act excessively aggressive or retreat and not engage in more normal social patterns."[50]

As with other childlike traits, human adults remain playful and trusting in a way that looks a lot more like Labradors than adult wolves or chimpanzees. When a grown wolf or a chimp bears its teeth, you'd better run. Humans, even adult humans, are by and large more into chasing balls than establishing dominance. The readiness with which we play with our friends and acquaintances and even strangers is remarkable, even though verbal banter or wordplay tends to gradually displace physical wrestling. When I joke with the hot dog vendor about his pathetic loyalty to the Mets, as evinced by the baseball cap he is wearing, we become very much like two dogs wrestling in a park: My verbal jabs are play-serious, not meant to genuinely wound, and the successful banter establishes an ephemeral but important trust connection in the midst of a busy metropolis. Insult a chimpanzee's favorite baseball team, on the other hand, and you're likely to lose an arm. The fact that humans retain into adulthood the complex and sophisticated cognitive machinery required to play, and in fact continue to enjoy playing with others, is a reflection of the profound importance of trust in human affairs.

REGAINING THE CHILD'S MIND

So, extended immaturity and the retention of childlike traits into adulthood can be seen as part of the human response to the challenges presented by the Three Cs. Through childhood and adolescence, we go through a long developmental stretch during which our minds wildly ping-pong from one thought to another. We are open to absorbing new information, and are both trusting and trustworthy, although less so as we age. Even as adults, though, to be successful in this extremely odd ecological niche we've carved out for ourselves, we need to remain creative, be able to absorb and transmit culture, and be capable of creating trust in others and solving cooperation dilemmas that require commitment. As a species, we are the Labradors of the primate world, retaining childlike traits into adulthood.

As countless myths and children's stories recount, however, childlike playfulness, something we uniquely crave among primates, is eventually lost. We relish some banter with the hot dog vendor, but keep it short because we're late for work. As adults, the childish drive to meander, examine boogers, and play becomes subordinated to productive routine. Get up, dress, commute, work, eat, sleep, repeat. This is the realm of the PFC, that center of executive control, and it is no accident that its maturation corresponds to an increased ability to stay on task, delay gratification, and subordinate emotions and desires to abstract reason and the achievement of practical goals.

And it couldn't be otherwise. Truth be told, as fun and endearing as they are, children are completely useless. If they were in charge of things we'd be doomed. My thirteen-year-old daughter cannot be relied upon to turn off the oven after she uses it, to remember to walk the dog, or even to hang up a wet towel rather than throw it in a heap on the floor. Even so, she is a laser-focused superachiever compared to her five-year-old self. The PFC is a physiologically expensive bit of machinery, and we evolved it for a reason. The ability to remain task-focused, repress emotions, and delay gratification is a crucial human trait. We cannot remain children forever.

This is why we shouldn't make too much of a four year-old's ability to outfox an adult in the counterintuitive version of the blicket test. Reflecting on their apparent creativity advantage, Alison Gopnik draws upon an analogy from the corporate world:

> There's a kind of evolutionary division of labor between children and adults. Children are the R&D department of the human species—the blue-sky guys, the brainstormers. Adults are production and marketing. They make discoveries, we implement them. They think up a million new ideas, mostly useless, and we take the three or four good ones and make them real.[51]

The problem with this analogy is that, in fact, very few patents have been awarded to four-year-olds. It is similarly hard to find examples of adult inventors directly borrowing ideas from their children.[52] Adults may sometimes take advantage of, or be inspired by, the random variation introduced by children messing about, but only if they are in a place to recognize useful innovation, capitalize on technical innovations and creative breakthroughs, or convert insights into products. At most, it is *virtual* youth—a childlike state of mind in an otherwise functional adult—that is the key to cultural innovation.

In order to successfully function as hive insects despite being apes, then, human adults need to be able to access childlike traits despite having fully developed PFCs. The goal is to temporarily regain the *mind* of a child, not to actually become a child again. We need to be creative and trusting, while also able to tie our shoes and get out of the house on time. Significantly, a common theme in cultures from across the world and throughout history is the idea of spiritual or moral perfection as somehow involving regaining the child's mind. The Gospel of Matthew declares, "Truly I tell you, unless you change and become like little children, you will never enter the kingdom of heaven." An early Chinese Daoist text, the *Daodejing* or *Laozi,* compares the perfected sage to an infant or small child, perfectly open and receptive to the world.[53]

In response to this need, humans have come up with various cultural technologies for temporarily but powerfully enhancing childlike creativity and receptiveness in otherwise fully functional adults. Spiritual practices of various sorts, such as meditation and prayer, can be effective ways to do it. Faster, simpler, and vastly more popular, however, is turning to chemical substances that can temporarily put development and cognitive maturation into reverse.

THE DRUNKEN MIND

As we saw in the creativity experiment described earlier in this chapter, if we want to re-create the cognitive flexibility of a child, a transcranial magnet would do the trick: We can just zap the PFC into submission. Such devices, however, have only become available recently. They are also expensive, not very portable, and typically not welcome at parties. What we need is something really low tech. Something that effectively takes the PFC offline and makes us happy and relaxed, but only for a few hours or so. Something that can be made anywhere, out of almost anything, by anyone, and produced reasonably cheaply. Bonus points if it tastes good, can be easily paired with food, and leads to dancing and other forms of communal sociality.

Alcohol, of course, fits these design specs to a T. The fact that it occurs naturally in rotten fruit means that its psychoactive properties are easily discoverable by a wide range of species. And what wonderful properties! On top of its ease of discoverability, production, and consumption, another factor contributing to alcohol's undisputed position as the king of intoxicants is its broad and complex range of effects on the human body and mind. As Stephen Braun observes, alcohol mimics the action of many other drugs, representing a kind of "pharmacy in a bottle: a stimulating/depressing/mood-altering drug that leaves practically no circuit or system of the brain untouched." It is unique in this respect among mind-altering drugs. "Substances such as cocaine and LSD work like pharmacological scalpels," Braun notes, "altering the functioning of only one or a handful of brain circuits. Alcohol is more like a pharmacological hand grenade. It affects practically everything around it."[54]

Partly this may be due to the ease with which alcohol spreads through our body-brain system. Ethanol, the active ingredient of alcoholic beverages, is soluble in both water and fat. Its water solubility means it is easily carried by water and absorbed quickly into the bloodstream, while its fat solubility allows it to easily cross fatty cell membranes.[55] While it is common to think of alcohol as a depressant, the real story is much more complicated, as one would expect from a pharmacological hand grenade.[56]

To begin with, alcoholic inebriation is "biphasal." The ascending phase, as blood alcohol levels rise, is characterized by stimulation and mild euphoria, as alcohol enhances the release of dopamine and serotonin. This is when alcohol mimics the effect of pure stimulants, like cocaine or MDMA. During this phase, alcohol also triggers the release of endorphins. You can think of it, in this regard, as something like a mild version of morphine, providing a pain-killing effect, enhancing overall mood, and decreasing anxiety.[57]

In its descending phase, as blood alcohol levels peak and then start to decline, alcohol has its depressant effect. In terms of inhibition of brain functioning, alcohol provides a depressant double-whammy. It enhances or exaggerates the activity of GABA$_A$ receptors, which function to inhibit neural activity, while simultaneously depressing the activity of glutamate receptors, which normally excite neural activity. So, when it comes to brain activity, alcohol simultaneously slams on the brakes and lets its foot off the accelerator. This neural screeching to a halt is what, especially at high blood alcohol levels, causes a sedative effect.[58]

The depressant effects of alcohol appear to be concentrated in three regions of the brain: the PFC, hippocampus, and cerebellum.[59] The hippocampus is implicated in memory and the cerebellum in basic motor skills. The fact that both are targeted by alcohol explains both why a drunken person might stumble and break a vase coming home after a night of revelry and then, in the morning, wonder how the vase got broken.

It is the downregulation of the PFC and associated regions, however, that is of the most interest to us. One of the crucial allies of the PFC when it comes to cognitive control is the anterior cingulate cortex (ACC). The ACC functions as a kind of playground monitor, watching your performance in the world and looking for errors or other negative feedback indicating that whatever you're currently doing should stop.[60] When you step out your door to walk to the office on a winter morning and begin slipping on the ice, it's the

ACC that notices this wobble and signals the PFC to take over from your motor system, which normally runs on autopilot. With the PFC in control, your walking is awkward and stilted, because motor systems run best when left to their own devices. It takes you longer to get to the office, but at least you don't slip and fall on your face.

In the lab, one way to see the ACC–PFC team in action is an experimental paradigm called the Wisconsin Card Sorting Test. In this task, subjects are given a set of cards with multiple shapes and symbols on them in various numbers and colors. They are then told to select the card that "matches" a stimulus card given to them by the experimenter. If they get it right, they're given positive feedback but no further instructions. It's a more sophisticated version of the color- or shape-matching task given to corvids or pigeons, and is really annoying at first because you are told when you succeed, but have no idea what the rules governing success might be. Nonetheless, subjects very quickly start to get the hang of it, because the matching principle is not random. It may be that you need to match the number of objects, or the color, or the type, but in any case, people zero in on the proper rule surprisingly quickly. Once they get comfortable with the rule they've intuited (say, match the shape and ignore number and color), and begin replying correctly by instinct, the nefarious experimenter, without telling them, changes the rule: Suddenly color matters and shape does not, so cards that matched before are now rejected. This is the laboratory equivalent of an icy patch of sidewalk suddenly not feeling the way you expect a sidewalk to feel. In neurologically typical subjects, the new errors caused by the rule change signal the ACC that something is wrong. The ACC then calls in the PFC to stop the previous behavior (matching by shape), slow down responses, and wait for a new rule to emerge. Once the new rule is figured out, the ACC is happy, the PFC can chill out, and the player can go back on autopilot, happily matching color to color.

This shift doesn't happen immediately. For some period of time after the rule change, people perseverate and continue pursuing the now-incorrect strategy in the face of negative feedback. The length of time it takes people to stop perseverating is a good measure of cognitive control and PFC health: For people with prefrontal damage or deficits, the change in circumstances takes a long time to alter behavior.[61] In other words, they continue to walk normally despite the new, icy conditions.

Significantly, drunk people perform on this task as if they had PFC damage. They continue to stubbornly plunge ahead in the face of negative feedback, which is not surprising to anyone who has watched a late-night pub returnee stupidly and repeatedly try to fit their house key into their neighbor's door.[62] In an excellent example of how alcohol's cocktail of effects reinforce one another, the impairment of the ability to recognize negative feedback is exacerbated by the various other explosions set off in the brain. The processing of fear and other negative emotions in the amygdala is depressed, so a drunk person becomes relatively insensitive to whatever negative stimuli manage to actually make it through.[63] Attention is narrowed to the immediate moment—so-called alcoholic "myopia"[64]—making it difficult to be swayed by abstract or external considerations and to anticipate future consequences. Working memory and cognitive processing speed are reduced.[65] The ability to inhibit impulses, one of the main jobs of the PFC, is impaired.[66] Finally, a person experiencing the serotonin and dopamine surge on the upswing phase of intoxication feels so good that even if their sluggish PFC and ACC team manage to get a warning signal through, they simply no longer *care* that they are messing things up.[67] This is the point at which the frustrated pub goer tosses their malfunctioning keys and breaks a window to get into "their" house, much to the consternation of their sleeping neighbors.

No other chemical intoxicant packs the broad wallop of alcohol, but the most popular have parallel effects on the human mind. The active component in cannabis, THC, targets its own receptors in the brain ("cannabinoid" receptors) in a manner that, like alcohol, boosts dopamine levels, interferes with memory formation, and impairs motor skills. Kava similarly appears to stimulate dopamine and reduce anxiety, but leaves most higher cognitive abilities relatively untouched. Classic hallucinogens, like LSD and psilocybin, target serotonin and possibly dopamine receptors, giving us the positive mood boost, while also seriously disrupting the brain's "default-mode network" (DMN). The DMN seems to provide us with our basic sense of self. Its disruption by hallucinogens creates radical cognitive fluidity, a fuzzy boundary between self and others, and a lack of sensory filtering that characterizes both the dreaming mind and the minds of small children.[68]

It is also worth noting that there are various non-chemical practices that can produce cognitive effects that mimic alcoholic intoxication. For instance, extreme exercise can produce a "runner's high" through stimulation of dopamine and downregulation of the PFC, as the overstressed body diverts resources away from the energy-hungry neocortex and toward more immediately necessary motor and circulation systems. The neuroscientist Arne Dietrich has argued that this combination seems to be responsible for the loss of sense of self and intense euphoria characteristic of athletic "peak experiences."[69] Various religious traditions have availed themselves of this trick. Sufi dancing, group singing and chanting, extended meditation in painful poses (cross-legged; kneeling in prayer), personal mortification (self-flagellation; piercing), or extreme breathing exercises can all provide a similar sort of high, boosting dopamine and endorphins while diverting energy away from the PFC.

What a hassle, though. Given the incredible amount of time and effort these extreme experiences require, it is no wonder that people turn to drugs. And among drugs, alcohol reigns supreme.

Cannabis needs to be smoked or ingested, is hard to dose, and has unpredictable effects on the body and mind.[70] Pot makes some people extroverted and energetic, others introverted, paranoid, and lethargic. Psilocybin mushrooms can produce intense gastro-intestinal distress, wild disorientation and delusions, and demand that you check out completely from reality for an extended period of time. This is why no culture encourages nibbling a few magic mushrooms before dinner or at a social reception. And compared to insanely toxic desert seeds or enormous poisonous toads, psilocybin is probably the most pedestrian and safe of naturally occurring hallucinogens, which isn't saying much.

Alcohol, on the other hand, is in many ways the perfect drug. It is easy to dose, and its cognitive effects are stable across individuals. Best of all, these effects wax and wane predictably and are relatively short-lived. Even as we happily accept that second drink from the bartender, our liver is furiously laboring to break down the ethanol from our first glass into something non-toxic so we can get it out of our body. So, given the downsides of most non-alcoholic chemical intoxicants, and the incredible amount of time, effort, and pain demanded by non-chemical means, it's not surprising that most of us, most of the time, have chosen to opt for a few pints of beer over sticking sharp objects through our cheeks. If we want to enhance mood and temporarily take the PFC offline, a delicious liquid neurotoxin seems to be the fastest and most pleasant option.

LEAVING THE DOOR OPEN FOR DIONYSUS

The PFC is the most evolutionarily novel part of the brain and the last to mature in development. It is also arguably what makes us human. It is difficult to imagine what life would be like without the ability to control impulses, focus on long-term tasks, reason abstractly, delay gratification, monitor our own functioning, and correct errors.

We have also seen, however, that when it comes to successfully responding to the demands of the Three Cs, to the particular challenges of occupying the human ecological niche, the PFC is the enemy. Intoxication is an antidote to cognitive control, a way to temporarily hamstring that opponent to creativity, cultural openness, and communal bonding.

To crystallize the relative benefits and costs of possessing this sober buzzkill in the front of our brains, let's once again draw on Greek mythology. The Greek pantheon includes two gods, Apollo and Dionysus, who personify the tension between self-control and abandon.[71] Apollo, the sun god, represents rationality, order, and self-control. In art, the Apollonian mode is characterized by restraint and elegance, carefully designed balance. Apollo was worshipped by means of staid, formal offerings in designated temples. Dionysus is the god of wine, drunkenness, fertility, emotionality, and chaos. Dionysian art indulges in excess, ecstatic elevation, altered states. His worshippers famously included the maenads, wild women who would gather secretly at night in the woods for the original *bacchanalia,* alcohol- and drug-fueled parties rather like contemporary raves, only with a much darker edge, more nudity, and occasional cannibalism.

Dionysus appeals to the more ancient, primitive regions of our brains, those dedicated to sex, emotions, movement, touch. Apollo finds his natural home in the prefrontal cortex. The PFC is what makes adult humans typically function more like grim wolves than playful Labradors. Under its guidance, we become very efficient at specialized tasks and are able to pursue them relentlessly, brushing aside boredom and distractions and fatigue. It is the source of everything that the four-year-old idly tying and untying her shoes so painfully lacks—at least, from the perspective of an adult trying to get to daycare and work on time. Caffeine and nicotine are the wolf's friends, helping her to focus, wiping away her fatigue, sharpening her attention. These substances are the friends and natural allies of the PFC. They are the tools of Apollo.

If, however, we want to give a leg up to our Dionysian nature, we need something that will slow down or impair the PFC, that will make us more playful, more creative, more emotional, more trusting. We need to become looser—one of Dionysus' alternative names in Latin was *Liber,* "The Free." We need something that will allow us to enjoy all of the wonderful qualities of the child-like mind as adults, to have our Apollonian order and discipline leavened with a bit of Dionysian chaos or relaxation.

This, of course, is why Dionysus is also the god of wine. Alcohol, ideally paired with a nice bit of music, dance, or other forms of play, is perfectly adapted for paralyzing the PFC for a few hours. After a glass or two, your attention is narrowed to only the immediate surroundings. You meander unpredictably, more free to follow wherever the conversation might take you. You feel happy and unconcerned about future consequences. Your motor skills are rubbish. On the other hand, if you speak a second language, you might find yourself suddenly a bit more confident and fluent. In other words, you are a child again, with all of the benefits and costs that come with stunting the PFC. This is an elegant and convenient solution to the problem of how to temporarily create a receptive, flexible, childlike state of mind in someone with the body, capabilities, and resources of an adult.

Allowing childlike Dionysus to take over, at least for a spell, is how we have responded to the challenges inherent to being human. Intoxication helps us with the demands of our ecological niche, making it easier for us to be creative, coexist in close quarters with others, keep up our spirits in collective undertakings, and be more open to connecting and learning from others. Even

Plato, an almost monomaniacal devotee of Apollo, recognized the need for the kind of mental and spiritual rejuvenation provided by alcohol: "The souls of the drinkers get red-hot, like glowing iron, and thus turn softer and more youthful, so that anyone who has the ability and skill to mold and educate them finds them as easy to handle as when they were young."[72] And getting drunk also helps us with the communal demands of being human, making us simultaneously more trusting and more trustworthy.

This is why alcohol use, despite its costs and the problems it brings in its wake, has not been eliminated by genetic evolution or cultural fiat. Whether literally or spiritually, from time to time, we *need* to get drunk. Apollo must be subordinated to Dionysus; the wolf needs to give way to the Labrador; the adult needs to cede her place to the child. In his seminal work on chemical intoxication, *The Doors of Perception*, Aldous Huxley wisely observes that "systematic reasoning is something we could not, as a species or as individuals, possibly do without. But neither, if we are to remain sane, can we possibly do without direct perception, the more unsystematic the better, of the inner and outer worlds into which we have been born."[73]

In other words, being human requires a careful balancing act between Apollo and Dionysus. We need to be able to tie our shoes, but also be occasionally distracted by the beautiful or interesting or new. Because of the distinctive adaptive challenges we face as a species, we require a way to inject controlled doses of chaos into our lives.[74] Apollo, the sober grown-up, can't be in charge all of the time. Dionysus, like a hapless toddler, may have trouble getting his shoes on, but he sometimes manages to stumble on novel solutions that Apollo would never see. Intoxication technologies, alcohol paramount among them, have historically been one way we have managed to leave the door open for Dionysus. And it is sipping, dancing, wildly ecstatic Dionysus who freed us from our selfish ape selves long enough to drag us, stumbling and laughing, into civilization.

CHAPTER THREE

INTOXICATION, ECSTASY, AND THE ORIGINS OF CIVILIZATION

Sometime around the eighth millennium BCE in the Fertile Crescent—a semi-circular swath of the Middle East spanning roughly present-day Egypt through Syria into Iraq and Iran—some clever hunter-gatherers started saving the seeds from unusually productive or tasty wild grains and legumes and replanting them in patches of clear land. They'd return the next season and do it again; maybe they'd also come back periodically to do a spot of weeding or watering. Eventually this process of selection produced early versions of familiar modern crops, plants productive enough to sustain and reward sedentary populations who could put in the work to clear land, sow and maintain the fields, and then hang around to reap the rewards. Voilà, agriculture. Somewhat later, similar processes independently led to agricultural revolutions in other regions of the world, such as the domestication of wheat, millet, and rice in the Yellow River and Yangzi River valleys in China and of maize (corn) in the Americas.

Once early farmers were systematically producing crops, they'd often end up with surpluses. These could be stored for the off-season, or as insurance against a future bad harvest. At some point, though, people noticed that if they left grain mashed up in water (say, as a result of an abandoned bread-making effort), the mixture would transform into something entirely different. It was not unpleasant smelling. It tasted a bit funny, but you got used to the flavor and even came to like it. Best of all, it could get you high. So, the story goes, sometime after mastering farming, humans also began to enjoy the benefits of beer, with similar processes around the world leading to grape, millet, rice, and maize-based alcoholic drinks. People finally had something tasty to pair with their bread and cheese. This is the standard account of the origin of alcohol production—that it's an accident, an unintended consequence of the invention of agriculture.

Around the 1950s, though, this story began to be questioned by proponents of various "beer before bread" theories.[1] They pointed out that large-scale, likely alcohol-fueled feasting, which often brought together people from far-flung regions for days of music, dancing, rituals, drinking, and sacrifices, began well before settled agriculture. At Göbekli Tepe, a site in what is now modern-day Turkey we'll talk about more below, hunter-gatherers convened regularly throughout the tenth to eighth millennia BCE to feast on gazelles, build circular structures, and erect enormous T-shaped limestone pillars carved with mysterious pictograms and animal forms—probably all while well-lubricated with beer. This is despite the fact that, practically speaking, getting drunk seems like a bad thing to combine with the construction of monumental architecture. The stone pillars erected at Göbekli Tepe weighed between ten and twenty metric tons and had to be transported almost half a kilometer from the site where they were quarried, requiring the hard labor of probably 500 people. It is easy to think that all of this cutting, dragging, and lifting was even more challenging when buzzed or with a hangover, not to mention the problem of inebriation-induced foot and finger crushing.

And yet, across the ancient world, we see similar evidence that the first large gatherings of people, centered on feasting, ritual, and booze, happened long before anyone had come up with the idea of planting and harvesting crops. Archaeologists working in the Fertile Crescent have noted that at the earliest known sites the particular tools being used and varieties of grain being grown were more suited to making beer than bread. One recent discovery found evidence of bread and/or beer making at a site in north-eastern Jordan dated to 14,400 years ago, predating the emergence of agriculture by at least 4,000 years.[2] Given that bread was still millennia away from becoming a dietary staple, the most likely motivation for these hunter-gatherers to hunker down and get to work was to produce the starring liquid ingredient of communal feasts and ecstatic religious rituals.[3] It is also worth noting that the world's oldest extant recipe is for beer—part of an early Sumerian myth—and that our earliest representations of group feasting include obvious depictions of alcohol swilling.[4] The human mastery of fermentation into alcohol is so ancient that certain yeast strains associated with wine and sake show evidence of having been domesticated 12,000 years ago or more.[5]

We see the same pattern of alcohol production preceding agriculture in other parts of the world. A primitive ancestor of maize called teosinte was cultivated in Central and South America almost 9,000 years ago, well before farmers managed to produce proper maize. Teosinte makes terrible corn flour but excellent booze; it's the basis for *chicha,* the beer-like beverage still drunk throughout Central and South America. There is evidence for *chicha* playing a role in ritualized feasting as early as the ninth century BCE, and its control and distribution was a crucial part of state ritual and power during the Inca Empire from the third century BCE.[6] The same is likely true of the drugs that took alcohol's place in areas that lacked it. In an echo of the beer before bread argument, some scholars have pointed to evidence that, in Australia, the desire to cultivate the ingredients for the intoxicant pituri drove the development of agriculture in certain regions.[7] Similarly, it is possible that the cultivation of tobacco in North and South America, especially in regions outside its native range, inspired the manipulation of other plant species and thereby the beginnings of agriculture.[8]

All of this suggests that it is quite likely that the desire to get drunk or high gave rise to agriculture, rather than the other way around. Agriculture, of course, is the foundation of civilization. This means that our taste for liquid or smokable neurotoxins, the most convenient means for taking the PFC offline, may have been the catalyst for settled agricultural life. Moreover, intoxicants not only lured us into civilization but, as we will explore in this chapter, also helped make it possible for us to become civilized. By causing humans to become, at least temporarily, more creative, cultural, and communal—to live like social insects, despite our ape nature—intoxicants provided the spark that allowed us to form truly large-scale groups, domesticate increasing numbers of plants and animals, accumulate new technologies, and thereby create the sprawling civilizations that have made us the dominant mega-fauna on the planet. In other words, it is Dionysus, with his skinful of wine and his seductive panpipes, who is the founder of civilization; Apollo just came along for the ride.

One of the many gifts attributed to Dionysus by the Greeks was the power of transformation. He could turn himself into an animal, and he was the god who granted the unfortunate King Midas the power to turn anything he touched into gold. As the god of intoxication, he could turn sane people mad. Or, even more impressively, he could transform task-focused, suspicious, aggressive, and fiercely independent primates into relaxed, creative, and trusting social beings. Let's now look at how, across the world and throughout history, humans have turned to Dionysus for help when confronting the challenge inherent to being a creative, cultural, and communal ape.

A VISIT FROM THE MUSE: INTOXICATION AND CREATIVITY

From the banks of the Ganges we heard
Of the god of joy's triumph,
As, from the Indus, the all-conquering Young Bacchus came,
with holy wine,
To rouse the people from sleep.
        —Friedrich Hölderlin, "The Poet's Vocation"[9]

A familiar trope in cultures around the world and throughout history is alcohol as muse. As Da'an Pan notes, "In traditional Chinese culture...wine plays the paradoxical role of intoxicator and facilitator of artistic imagination, 'awakening' drinkers to their optimum creative moments...to be intoxicated is to be inspired."[10] It is not uncommon for ancient Chinese poets to have entire series of poems under the rubric, "Written While Drunk," including this one from the Zhang Yue (667 to 730):

Once drunk, my delight knows no limits—
Even better than before I'm drunk.
My movements, my expressions, all turn into dance,
And every word out of my mouth turns into a poem![11]

This recalls an ancient Greek saying:

If with water you fill up your glasses
You'll never write anything wise
But wine is the horse of Parnassus
That carries a bard to the skies.[12]

The name of Kvasir, the Anglo-Saxon god of the bards and artistic inspiration, was derived from the word for "strong ale." In Norse mythology, Kvasir is killed and his blood mixed with alcohol to create the "mead of inspiration." "Anyone who drank of this magic potion," observes Iain Gately, "could thereafter compose poetry and speak wise words."[13] And despite a nominal religious commitment to prohibition, the greatest Persian poetry was produced by, and openly celebrated, the revelatory power of wine.

As the scholar of English literature Marty Roth notes, while modern writers from Eugene O'Neill to Hemingway have explicitly denied the role of alcohol in their art, "this disclaimer, when it comes from a heavy drinker, is more likely to be part of an alcoholic alibi system than a statement of fact."[14] In any case, it is impossible to ignore the fact that an inordinate proportion of writers, poets, artists, and musicians are also heavy users of liquid inspiration, willing to put up with the physical and sometimes financial and personal costs in return for an unleashed mind. "It shrinks my liver, doesn't it?...It pickles my kidneys," declares the fictional, and alcoholic, writer in Billy Wilder's *The Lost Weekend*. "Yes. What does it do to my mind? It tosses the sandbags overboard so the balloon can soar."[15] Other chemical intoxicants can also be drafted to play this role. In Vanuatu, a Pacific island that traditionally relied on kava rather than alcohol, songwriters, when commissioned to create something new, withdraw to the forest to commune with the ancestors, drink kava, and await inspiration.[16] Cannabis has similarly served as a muse for millennia, fueling the imaginations of Sufi mystics, Beat poets, and jazz musicians.[17] And of course Coleridge's famous "Kubla Khan" was the product of an intense opium trip.

The advantage of alcohol, kava, and cannabis is that they are relatively easy to integrate into ordinary creative and social life. Psychedelics like psilocybin or mescaline create a more dramatic rupture with everyday reality. For this reason, their use has historically been restricted to occasional rituals or to a specialized social class, such as shamans. The term "shaman" is generally used by scholars of religion to refer to individuals who serve as intermediaries between the human and spirit worlds, and derive from the latter various powers, typically including the ability to cure illnesses, foretell the future, and communicate with and control animals.[18] Across the world, some of our earliest unearthed burials involve what seem to be the remains of shamanistic figures, based on the presence of ritual items, animal imagery, and—perhaps the most commonly cited diagnostic feature—chemical intoxicants, particularly psychedelics.

So ancient are shamanistic religions that some claim they can be found even among other, extinct hominid lines. The so-called "flower burial" in a cave in northern Iraq, from approximately 60,000 years ago, contains the remains of a Neanderthal (*Homo neanderthalensis*) male who was described in early reports as a possible shaman, based on pollen traces suggesting that he had been laid to rest on a bed of flowers that included a wide range of medicinal and intoxicating drugs.[19] Whatever the likelihood that Neanderthals got high or communed with spirits, early human burials suggest that intoxicant-fueled shamanistic trances and visions have an ancient pedigree. Ceramic vessels from the Chavín culture in the Peruvian Andes (1200 to 600 BCE) portraying a jaguar beneath a San Pedro (mescaline) cactus are generally understood in a shamanistic context. The cactus is the drug that facilitates the shaman's visions and journeys to the other world. In the Andes, a cave burial dated to approximately 1000 CE appears to be the tomb of a shamanistic figure, given the presence of extensive ritual paraphernalia, snuffing tables and a snuffing tube, and a pouch containing the remains of cocaine, chemicals associated with ayahuasca, and possibly psilocybin. None of these substances are indigenous to the region, suggesting that intoxicants were part of an established and wide-ranging trade network.

One way to look at the historic function of shamans is as sources of radically creative insights. Returning from their chemically in-duced spirit journeys, they bring with them new views about what is making So-and-So sick, the root causes of the conflict between factions X and Y, or what to do about the recent drought or disappearance of local game. While shamans typically credit these insights to the spirits, we might see them, instead, as messages from deep regions of the unconscious allowed to rise to the surface of a PFC-impaired brain. With the benefit of modern cognitive science, we can understand that this ancient and ubiquitous association of intoxication and creative insight is no accident.

Psychologists have long known that the sort of creativity involved in lateral thinking is inhibited by an explicit focus on goal achieve-ment or concern about rewards external to the task at hand. Trying too hard to solve the Remote Associate Task (RAT) makes you do worse than if you simply relax and give your mind space to come up with a key word that unites the three stimulus words.[20] When instructed to produce a creative collage from a stack of colored paper and glue, you will produce a more dull and conservative effort if you think you are to be evaluated by external judges.[21] Although people with reduced cognitive control capacity have trouble focusing their attention, they appear to do better at solving problems that require creativity and flexibility.[22] Professional writ-ers and physicists alike report that their most creative ideas—the ones that produce the wonderful feeling of *aha!* discovery—come to them while "mind wandering," or drifting around in a mental state disconnected from the task at hand or immediate goals.[23]

As we have seen earlier, it is the PFC, the seat of Apollo, that is the problem here. We noted that adults with PFC damage, or those who have their PFCs temporarily taken offline by a nice zap from a transcranial magnet, do better on creativity tasks. Similarly helpful is an overall passive or relaxed state of mind, indicated by a high level of alpha-wave activity in the brain, which reflects a downregulation of goal-oriented and top-down control regions like the PFC. In one study, experimenters used biofeedback to increase alpha-wave activity in a group of subjects.[24] The participants were wired up to EEG monitors, shown a screen with a green bar indicating their level of alpha activity, and instructed to raise the bar as high as possible. To help them, they were given hints that would be familiar to anyone who has tried meditation: Relax your mind, breathe deeply and regularly, let all thoughts and feelings come and go freely, feel your body relaxing into your posture. Shortly thereafter, the subjects who successfully raised their alpha activity outperformed their peers on a lateral thinking task.

As the creative primate, humans are crucially dependent on lateral thinking. We require a continuous stream of novel insights and a constant reorganization of existing knowledge. Children, with their underdeveloped PFCs, are superstars in this regard. But as we've seen, the very thing that makes them so creative renders most of their creations useless, at least from the pragmatic perspective of goal-oriented adults. Bizarrely distorted Lego worlds featuring post-apocalyptic, scavenged-parts vehicles driven by Lego people with Barbie-doll heads, or menageries of superhero figurines and stuffies organized into formal English tea parties, reflect impressive out-of-the-box thinking. But what society really needs right now is new vaccines and more efficient lithium-ion batteries. If your goal is to maximize implementable cultural innovation, your ideal person would be someone with the body of an adult but, for a brief period, the mind of a child. Someone with downregulated cognitive control, heightened openness to experience, and a mind prone to wander off in unpredictable directions. In other words, a drunk, stoned, or tripping adult. Societies have come to associate intoxication with creativity because chemical intoxication has been a crucial and widely used technology to effect this transition from adult to mental childhood in a relatively controlled manner.

CHEMICAL PUPPIES: TURNING WOLVES INTO LABRADORS

Leo Tolstoy was an uncompromising hard-ass when it came to confronting reality. He therefore had a predictably dour view of intoxicant use. In his 1890 essay, "Why Do Men Stupefy Themselves?," he declares that "the cause of the worldwide consumption of hashish, opium, wine and tobacco lies not in the taste, nor in any pleasure, recreation, or mirth they afford, but simply in man's need to hide from himself the demands of conscience." Ouch.

It is certainly the case that the characters in Tolstoy's novels use intoxicants to numb misgivings about their moral transgressions or dissolute lifestyles. And yet it is not hard to find defenses of alcohol as a source of genuine joy and comfort in the religious traditions of the world. An ancient Sumerian hymn to the Beer Goddess declares:

> Let the heart of the fermenting vat be our heart!
> What makes your heart feel wonderful,
> Makes also our heart feel wonderful.
> Our liver is happy, our heart is joyful.
> You poured a libation over the brick of destiny...
> Drinking beer, in a blissful mood,
> Drinking liquor, feeling exhilarated,
> With joy in the heart and a happy liver.[25]

The joy provided by alcohol is often linked to a kind of mental evasion, as Tolstoy worried, but typically what is being fled is not one's conscience but rather the harshness of everyday life. As we read in the Book of Proverbs, "Give strong drink unto him that is ready to perish, and wine unto them that be of heavy hearts. Let him drink, and forget his poverty, and remember his misery no more."[26] Poets certainly find not only inspiration but also emotional comfort at the bottom of a glass. "We must not let our spirits give way to grief," writes the Greek poet Alcaeus. "Best of all defenses is to mix plenty of wine, and drink it."[27] The Chinese poet Tao Yuanming declares:

A myriad tribulations follow one another
Is human life not exhausting?
How can I satisfy (*cheng* 稱) my emotions?
With cloudy wine just let me enjoy myself.[28]

In the context of this poem the verb *cheng* is probably best rendered as "satisfy," but the root meaning is to weigh on a scale or to adjust or harmonize a system. As in the Book of Proverbs quotation, we get a clear sense of alcohol being used as a tool for mood regulation.

If even relatively well-off elites like Tao Yuanming and Alcaeus find the need to drown their sorrows or indulge in a moment of reality-denying joy, imagine how intense the need has been for the vast majority of those who have lived in the large-scale societies of the world, toiling away in the fields, workshops, roads, and building sites, scraping by day to day on inadequate food and rest. For people like this, a two- to three-hour vacation from reality is not only pleasurable but perhaps necessary.

Stress or anxiety reduction has been the most common social function attributed to alcohol by anthropologists inclined to try to explain intoxicant use in the first place.[29] Perhaps the most prominent early proponent of this view was Donald Horton. In a survey of drinking practices across fifty-six small-scale societies published in 1943, Horton declared that "the primary function of alcoholic beverages in all societies is the reduction of anxiety."[30] He proposed a hydraulic model of alcohol use, arguing that the rate of drinking rises along with an increase in anxiety-generating food scarcity or war, until it runs up against new anxieties generated by excessive drinking. Any given society ends up at an equilibrium between these two extremes.

Horton's specific theory has not aged well. It remains the case, however, that alcohol's ability to boost overall mood and relieve anxiety and stress remain at the heart of both folk and scientific explanations of intoxicant use.[31] What has changed is a recognition that mood boosting and anxiety reduction at the individual level probably serve a broader social function of enabling humans to rub along better in the cramped, hierarchical confines of large-scale societies. More recent anthropological theories see alcohol as a tool for enhancing social solidarity, allowing fiercely independent hunter-gatherers to shrug off those aspects of their chimpanzee nature that present a barrier to living like a social insect.[32]

To get a sense of how stress relief is also fundamentally a social problem, rats can be a helpful model. A study examining the relationship between stress and voluntary alcohol consumption in rats subjected three groups of healthy rats, who had never tasted alcohol before, to differing amounts of stress.[33] A control group was placed in a normal, uncrowded laboratory cage free of daily stressors. An acutely stressed group spent six hours a day in a tiny, overcrowded cage where they could barely move, but spent the rest of their time in a normal cage. A third, chronically stressed group spent the entire weeklong study period in a less dramatically overcrowded, but still uncomfortably close, cage. All of the rats were allowed free access to food and two liquid sources: a bottle containing only tap water and one containing tap water heavily spiked with ethanol.

Compared to the control group, the acutely and chronically stressed groups lost body weight over the course of the study, a sign that they were genuinely freaked out by their conditions. Both groups responded to this stress by hitting the bottle: after only one day of overcrowding, their consumption of alcohol was significantly more than the control group. This is interesting, but perhaps not all that surprising. The ways in which the behavior of the acutely and chronically stressed groups diverged, however, is more revealing. Alcohol consumption by the acute group remained fairly stable, and by the end of the week was more or less matched by the control group, who had apparently also developed a mild interest in this novel beverage. The chronically stressed rats, on the other hand, turned to booze with a passion, ramping their alcohol consumption up to levels that would make F. Scott Fitzgerald blush (Figure 3.1).

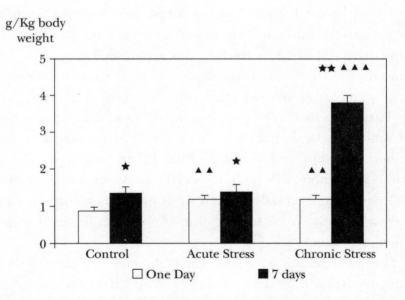

Figure 3.1. One-day and seven-day alcohol consumption levels for control, acutely stressed, and chronically stressed rats.[34]

The authors of the study concluded that temporary stressors can be easily adapted to without chemical help, whereas long-term stress motivates the sustained consumption of liquid comfort in quantities that need to be ramped up over time to maintain the effect.

This study seems like a pretty cruel thing to do to rats. It is arguable, however, that our species' transition from small hunter-gatherer bands—the lifestyle we enjoyed for most of our evolutionary history, and the one we share with our closest relatives and immediate ancestors—to settled agricultural communities involved a similarly brutal shock.

Greg Wadley and Brian Hayden, recent and prominent proponents of the beer before bread hypothesis, argue that the Neolithic transition to agriculture seriously increased both crowding and inequality. Hunter-gatherer bands likely consisted of twenty to forty people roaming across a broad landscape in search of game and plants. Those who lived through the lifestyle revolution that first occurred in the Fertile Crescent, when mobile hunter-gatherers began to settle into much larger and more sedentary communities, must have felt like rats thrown into a too-small cage with pretty crummy provisions. It certainly involved a marked decrease in quality and variety of food, from a diverse mix of wild meats, plants, and fruits to a diet based on filling but dull and vitamin-poor bread or other starches. There was also a steady and dramatic increase in both crowding and inequality. Even 12,000 years ago, as Wadley and Hayden note, villages in the Fertile Crescent contained 200 to 300 people and already showed signs of private property, wealth inequality, and social stratification. After that, things got much worse, very quickly.

Empirical studies have shown that the effect of alcohol on the human response to stress is similar to its effect on that of rats, as might be expected from our survey of alcohol's physiological effects in Chapter Two. In one classic study,[35] male volunteers from the Indiana University community were subjected to arguably more intense stressors than overcrowding in a cage: They were made to watch a digital clock countdown from 360 to 0, at which point they would receive a painful electric shock or have to give an extemporaneous speech, into a camera, on the topic, "What I like and dislike about my physical appearance." This speech would then be rated by a panel of judges for degree of openness and level of neuroticism. (Human subject approval was apparently much easier to get in the 1980s.) Their stress response was continuously monitored with implicit measurements, like heart rate and skin conductance (which increases with stress), as well as explicit reports provided by an "anxiety dial." This was something subjects were instructed to turn to reflect their conscious mental state, ranging from 1 ("extremely calm") to 10 ("extremely tense"). Their blood pressure, self-reported mood, and blood alcohol content (BAC) were also assessed before and after the experiment.

To control for "expectancy effects," or cultural views about what alcohol should do to you, experimenters used what is known as a balanced placebo design. Everyone was given what looked and tasted like a vodka tonic with lime juice, but groups were told different things about what they were drinking. The first group was told they were getting a vodka tonic, and they did; the second group was told the same thing, but they were given plain tonic water instead. (At the cocktail strengths used in the study, the alcoholic and virgin versions were demonstrated to be indistinguishable by taste.) A third group was told they were merely getting tonic, but got a drink spiked with vodka instead. Finally, a fourth group was told they were getting a virgin cocktail, and that's precisely what they got. Subjects in the two alcohol conditions (groups #1 and #3) ended up with BACs around .09 percent, just over the legal limit for driving. During the experiment they displayed lower skin conductance and heart-rate increases, as well as lower self-reported anxiety levels. After the experiment they reported being more cheerful than controls. Significantly, the effects of alcohol were entirely due to the pharmacological properties of ethanol: Subjects in group #2 showed no placebo effects, either physiological or psychological.

This study from the 1980s has since been bolstered by a massive literature demonstrating alcohol's "stress-response dampening" effect.[36] Mild levels of inebriation reduce our physiological and psychological response to a wide variety of stressors, both physical (loud noises, electric shocks) and social (public speaking, conversations with strangers). This calming effect stems from alcohol's complicated and broad spectrum of effects on the human body, encompassing both its stimulating (increased energy, mild euphoria) and depressing (relaxation, muscle-tension reduction, cognitive myopia) functions.[37]

Turning back to rats, stress from overcrowding isn't the only thing that drives them to drink. They also hit the bottle in response to being defeated in social interactions. Subordinate mice chronically housed with more dominant and territorial males drank significantly more alcohol than control peers allowed to live placidly on their own, and also upped their alcohol consumption in the wake of being bullied by the dominant males.[38] As Wadley and Hayden dryly observe, "Heavy use in humans is likely due to similar situations." It may very well be the case that, as discussed in Chapter One, the "biological ennoblement" of vitamin- and trace mineral–poor grains through fermentation was a boon to early farmers trying to subsist on a monotonous diet. Far more important, however, were the psychological effects, the "joyful heart and happy liver" bestowed by the goddess of beer, who eases suffering by "pouring a libation over the brick of destiny."

A wonderful illustration of this social function of alcohol is provided by the movie *Babette's Feast* (1987). It centers on a small, isolated community of dour pietists on a remote portion of the Danish coast. They live a spare, ascetic, and completely dry lifestyle, eschewing alcohol and other intoxicants. Their primary joy is regular and mildly ecstatic (for Danish people, at least) religious ceremonies, which are organized and presided over by their charismatic leader. Once this leader dies, these services become less cohesive, and the community begins to fragment. Old personal grievances are revived, past slights remembered, and church gatherings begin to look more like a grudging collection of chimpanzees than the harmonious coordination of bees. What ultimately restores harmony to the community is a night of baccha-nalia, orchestrated by an outsider, Babette, a French chef who'd been forced to flee revolutionary Paris. Babette's feast provides a dazzling array of elaborate and exotic food, but, we cannot help noticing, it is driven and lubricated by a relentless stream of imported, world-class alcohol. Over the course of the evening, tensions relax, jokes fly, and old friendships are restored. It is hard to imagine a better fictional portrayal of the reduction of stress and interpersonal conflict that accompanies a steady and sustained rise in a group's collective BAC. There are many ways humans can achieve a hive mind, but liquor is certainly the quickest.

It is therefore no accident that, in the race to cut back on airline service that led from the full, complimentary meals of my childhood to the stingily doled out packets of crackers in today's economy class, one thing that was never dropped was the avail-ability of alcohol. One crucial way humans have succeeded in not tearing each other limb from limb when crammed into close quar-ters, or forced to work in subordination to others, is to consume just the right amount of a mild intoxicant. Nowadays we tend to reserve chemical stress alleviation for the end of the workday, relaxing with a drink or two at home or at the pub. Our ancestors, on the other hand, generally took the edge off with beers that were quite weak by contemporary standards, and spaced out their self-medication over the entire course of the workday. In any case, if the beer before bread advocates are correct, alcohol not only drove the creation of civilization by motivating early farmers to settle down and produce grain to ferment, but also by providing them with an invaluable tool to manage the psychological stress that came with this dramatic change in lifestyle.[39]

THE CHEMICAL HANDSHAKE: *IN VINO VERITAS*

In the last chapter we talked about the pervasiveness of trust-based relationships in human affairs, and how emotional commitment can allow us to solve otherwise intractable cooperation dilemmas. Learning to trust others is crucial for the communal primate. One thing we did not mention there, however, is that commitment relationships, despite their obvious benefits, are vulnerable to a unique form of defection: hypocrisy. If I can fake a commitment to you—making a big show of tying myself to the mast like Odysseus, but leaving the knots undone—I can reap all of the benefits of commitment without paying any of the costs. In terms of the Prisoner's Dilemma, the cooperation challenge we discussed there, I get off with a one-month sentence while you rot in jail for three years. You help me move my couch, but I've got a mysterious back injury or a flat tire when it's your turn to move.

In order to enjoy the benefits of being sincerely communal, humans therefore have had to learn how to trust, but not indiscriminately. This need has driven the evolution of various capacities for evaluating the sincerity and trustworthiness of others, including reading micro-facial expressions, voice tone, and body language.[40] Research shows that we size up and evaluate the trustworthiness of others almost immediately upon meeting them. One study found that subjects judged the trustworthiness of faces within 100 milliseconds, and that these judgments did not change even when people were given more information or time.[41] This tendency to instantly peg people as likely cooperators or not appears quite early in development, with children above the age of three quickly and readily classing faces as "mean" or "nice."[42] These gut-level assessments are consistent across cultures, and play a surprisingly outsized role even in formal contexts, like court cases or political elections, where you would expect people to be guided by more abstract and rational criteria.

The gut check is also important in negotiating cooperation challenges in public goods games like the Prisoner's Dilemma. In one experiment,[43] pairs of strangers were allowed to interact with one another for thirty minutes before playing a one-shot Prisoner's Dilemma game that provided a real incentive to cheat. Cooperative pairs were quick to accurately suss each other out and went on to solve the dilemma, getting the best overall payout. Interestingly, defectors were also able to identify fellow defection-prone partners and withhold cooperation—there were very few cases of mismatch, where a defector managed to take advantage of a cooperator. This result has been successfully replicated many times, and accurate predictions seem to be based on implicit cues derived from facial expressions, body language, and tone of voice.[44]

When it comes to deciding whether or not to trust someone, we prefer to rely on cues like emotional expression and subtle body language because they are relatively independent of conscious control. We know, at least implicitly, that the PFC, the locus of cold calculation and self-interest, is something we need to keep our eye on, so we're intuitively more likely to base our evaluations of trustworthiness on signals that bypass its control. Emotions tap right into our hot, unconscious cognition. They tend to arise without warning and are very difficult to control, as anyone who tried to suppress a genuine smile or look of horror knows well.

The idea of facial displays of emotion as honest, hard-to-fake signals goes all the way back to Charles Darwin in the modern scientific literature, although this communicative function of emotions was well-known to ancient thinkers in China and Greece.[45] People can quickly and accurately recognize and classify emotions observed in others,[46] and the suppression of emotional "leakage" in facial expressions or tone of voice is difficult to effectively pull off.[47] We are adept at distinguishing genuine smiles and spontaneous laughter from forced smiles and laughter. In fact, the two types of displays involve different muscular and vocal systems,[48] with the former being much less subject to conscious control. In real-life public goods games, people are more trusting—risking larger stakes and therefore benefiting more—when interacting with partners who displayed genuine as opposed to forced smiles.[49] Following a defection, people are more likely to forgive and trust a repentant partner who *blushes*, blushing being a classic involuntary reaction.[50]

In one disturbing but elegant study,[51] the psychologist Leanne ten Brinke and colleagues coded video clips, taken from real life, of individuals emotionally pleading to the public for information that would help with the return of a missing relative. In half of these cases, the pleaders were lying, and had been later convicted, based on overwhelming physical evidence, of having murdered the relative in question. Unaware of which people were later found to have been faking their distress, participants were nonetheless able to pick them out by focusing on facial muscle groups that are difficult to consciously control. The murderers showed less activation of genuine "grief muscles" (corrugator supercilii, depressor anguli oris) and more activation of muscles associated with fake smiles (zygomatic major) and conscious attempts to appear sad (frontalis).

Trustworthiness, then, is linked in our minds to the perceived authenticity and spontaneity of emotional displays.[52] This makes sense. We distrust people who seem emotionless or insincerely emotional—in the terms of our analogy to Odysseus and the sirens, these people haven't tied themselves to the mast, or have done so only loosely. This bias against Mr. Spock in favor of Captain Kirk is validated by recent experimental work suggesting that people are more cooperative in public goods games when forced to decide quickly or told to trust their intuition.[53] Telling them to reflect, or forcing them to take their time in making a decision, brings out the rational sneak and generates more cheating at the expense of the public good. There is a good reason religious and ethical systems across the world and throughout history have linked spontaneity and authenticity to moral reliability and social charisma.[54]

Our ability to weed out hypocrites is thus a crucial part of communal human life, and we are pretty good at detecting sketchy prospective partners. So, it seems that humans have solved this central danger at the heart of our communal life—the risk of hypocrites free-riding on commitment—through a clever evolutionary trick: paying attention to non-conscious, emotional signals. Use them to size up potential partners, and simply walk away from those who strike you as shifty.

Evolution, unfortunately, never rests. As lions get faster and better able to catch gazelles, only the fastest gazelles survive, slowly driving up the speed of gazelles. Now only the fastest lions manage to catch their prey, creating renewed pressure for speed on the part of lions. And so on. These sorts of evolutionary arms races can be seen throughout the biological world,[55] and are typically the engine driving the development of extreme traits.

One such trait is precisely this human ability to ferret out dishonesty. Although we take it for granted that we can immediately tell that that hot dog vendor seems a bit shifty, or that our child is lying about having walked the dog, a chimpanzee would be astounded by our mind-reading capacities—this would all seem like magic to them. Chimps seem capable of rudimentary mind-state signaling,[56] but our ability to transmit an enormous bandwidth of thoughts, emotions, and character traits to one another through a slight raise of an eyebrow, tone of voice, or twitch of the mouth is absolutely unmatched in the animal world. It bears all of the hallmarks of being an extreme trait driven by an evolutionary arms race.

When we see super-fast antelopes darting across the plains of North America, we infer the presence of almost-as-super-fast predators who motivated this speed—in the case of American antelopes, actually the "ghosts" of predators, like lions and cheetahs, who went extinct in the region thousands of years ago.[57] Our seemingly supernatural ability to detect lies has similarly been driven by a corresponding ability to deceive. Humans are world-class liars, and we've been getting better at it for millennia. Particularly fast contemporary antelopes are successful, at least in part, because they are able to voluntarily control muscle systems not normally under conscious control. The actor and director Woody Allen, for instance, belongs to a small group of people who can control a set of forehead muscles that allows him to make that trademark "I know you think I did something bad or that I'm a schmuck, but I'm really just misunderstood" expression. Being able to produce that face on command, especially when others can only produce it spontaneously, really comes in handy.[58] Charismatic politicians, like Bill Clinton, seem capable of temporarily but genuinely convincing themselves that their current interlocutor, maybe a small businessperson concerned about tariffs, is the only person in the entire world they care about, and to give them what appears to be their full attention, even while part of their brain might be intent on the big donor across the room. There is some evidence that psychopaths, the ultimate social defectors, are able to suppress genuine emotional "leakage,"[59] contributing to their creepy ability to stone-cold lie.

On the lion's side, fortune-tellers probably represent the leading edge of cheater detectors. When they hold your hand, look into your eyes, and ask you vague then increasingly precise questions, they are taking advantage of subtle expressions and tiny reactions to hone in on the fact that you recently suffered a loss in your family and are desperately unhappy at work. This seems like magic to ordinary people, but only to the same extent that common-place human mind-reading would seem to a chimpanzee. The science fiction writer Arthur C. Clarke once said that any sufficiently advanced technology is indistinguishable from magic to the uninitiated. The same is true of extreme traits exaggerated by runaway evolutionary arms races.

Crucially for our story about alcohol, cultures are not disinterested bystanders in this contest between cheaters and cheater detectors. Cultures benefit when the individuals in them are able to solve Prisoner's Dilemmas and other cooperation challenges, and so have a vested interest in putting their thumb on the scale to favor the cheater detectors. They do this by targeting a chink in the cheater's armor, the fact that cheating or lying requires cognitive control. It's easy and effortless to appear to be honest or sincere when you are telling the truth or expressing a genuine emotion; formulating a lie or faking an emotion requires effort and attention. If you want to make it harder for liars to lie, one promising approach would be to exploit this weakness by downregulating their cognitive control. Ideally, you'd want to do this in any important social situation where cheating might be a concern, and in an unobtrusive manner. No transcranial magnets allowed. Bonus points if you can do it in a way that is actually pleasurable and also makes people happy and more focused on those around them.

You see where I am going with this. I've spent so much time

here on the evolutionary dynamics of commitment and cheater detection because the threat presented by the hypocrite, the false friend, is an existential one for any community. This is why helping to unmask fakers, and thereby solidify interpersonal trust, is a crucial function that intoxicants have played in human civilization.[60] There is a very good reason that, in societies as different as those in ancient Greece, ancient China, medieval Europe, and the prehistoric Pacific Islands, no gathering of potentially hostile individuals occurred without the inclusion of staggering quantities of intoxicants.

A recently discovered ancient Chinese text, dating to the fourth or third century BCE and written on bamboo strips, contains the evocative declaration, "Harmony between states is brought about through the drinking of wine."[61] In ancient China no political agreement was reached without the participants first voluntarily impairing their brains with carefully timed and calibrated shots of liquid neurotoxin. The Roman historian Tacitus noted that, among the barbarian tribes of Germany, every political or military decision had to be run through the gauntlet of drunken communal opinion:

> It is at their feasts that they generally consult on the reconciliation of enemies, on the forming of matrimonial alliances, on the choice of chiefs, finally even on peace and war, for they think that at no time is the mind more open to simplicity of purpose or more warmed to noble aspirations. A race without either natural or acquired cunning, they disclose their hidden thoughts in the freedom of the festivity. Thus the sentiments of all having been discovered and laid bare, the discussion is renewed on the following day...They deliberate when they have no power to dissemble; they resolve when error is impossible.[62]

Although Tacitus patronizingly portrays this truth-serum use of alcohol as a primitive, barbarian practice, the ancient Romans and Greeks themselves relied heavily on precisely these functions. Indeed, the idea that drunkenness reveals the "true" self, though ancient and universal, is perhaps most famously expressed by the Latin *in vino veritas*, "in wine there is truth." This perceived link between honesty and drunkenness goes back to the Greeks, for whom the combination of "wine and truth" was a truism. "Inappropriate sobriety was thought highly suspect," Iain Gately notes. "Some skills, such as oratory, could only be exercised when drunk. Sober people were coldhearted—they meditated before they spoke and were careful about what they said, and therefore...did not really care about their subject."[63] A line in Plato's *Symposium* declares that "truth is revealed by wine and children"—a very telling equation of the sort of PFC impairment shared by children and drunks.

Since drunken words are spoken straight from the heart, they have historically been accorded greater weight than communications from the sneaky, controlled, and calculating self. In ancient Greece, oaths declared under the influence of wine were viewed as particularly sacred, reliable, and powerful. The Vikings similarly accorded an almost magical reverence to vows made after drinking (heavily) from a sacred "promise cup"; in Elizabethan and Stuart England, public declarations were viewed with suspicion unless they were accompanied by toasting with alcohol.[64]

My favorite illustration of the truth- and trust-enhancing function of alcohol is fictional, from the TV show *Game of Thrones*. In the famous "Red Wedding" episode, two rival clans have apparently overcome their differences and agreed to unite against their common enemy. As humans are wont to do, this agreement to cooperate is celebrated and reinforced by an alcohol-soaked, wildly drunken banquet. In the midst of the revelry, a servant begins to pour more wine for Lord Bolton, a decidedly shifty character, but he puts his hand over the glass. When his drunken neighbor asks him, incredulously, why he isn't drinking, Bolton replies tersely, "Dulls the senses." "That's the point!" is the cheerful reply. Indeed, that is the point. As any *Game of Thrones* fan is aware, Lord Bolton is a classic defector, keeping his mind clear so that he can direct the coldhearted murder of all of his drunken "friends." The take-home lesson is: Keep your eye on the guy who is skipping the toasts.

Alcohol is the most commonly used truth-telling technology, but it is revealing that, in regions without alcohol, other intoxicants play an identical functional role. The earliest European explorers in the Pacific reported being welcomed, and also having their degree of threat assessed, through kava-centered banquets.[65] To this day, no Fijian village counsel can begin deliberations until all present are appropriately high on kava. Similarly, among Woodlands and Plains tribes in North America, rival chiefs settled disputes and concluded conflicts over the calumet or "peace pipe" later celebrated in Hollywood Westerns. What was conspicuously left out of these cinematic re-creations was the intensely intoxicating effect of these hallucinogen-laced smokes. "Custom dictated that if the calumet was offered and accepted, the act of smoking would make any engagements sacred and inviolable," notes the historian of American religion Robert Fuller. "It was thought that anyone who violated this agreement could never escape just punishment."[66]

The fact that, when alcohol is taken out of the equation, other chemical intoxicants are tapped to fill the same functional role is a strong bit of evidence against any hijack or hangover theory. Although they did not enjoy the benefits of modern neuroscience or social psychology, cultures throughout time and across the world implicitly understood that the sober, rational, calculating individual mind is a barrier to social trust. This is why it is common for drunkenness—often serious drunkenness—to be obligatory for important social occasions, business negotiations, and religious rituals. An ancient Chinese poem from the *Book of Odes* declares:

> Sopping lies the dew;
> Not till the sun comes will it dry.
> Deep we quaff at our night-drinking;
> Not till we are drunk shall we go home.[67]

The Jewish holiday of Purim, honoring Mordecai's victory over the genocidal Haman, similarly demands that the celebrant become so inebriated that he or she cannot tell the difference between "Cursed be Haman" and "Blessed be Mordecai."

Just as we shake hands to show that we are not carrying a physical weapon, communal intoxication allows us to cognitively disarm in the presence of others. By the tenth toast of sorghum liquor at a Chinese banquet, or the final round of wine at a Greek symposium, or the end of Purim, the attendees have all effectively laid their PFCs on the table, exposing themselves as cognitively defenseless. This is the social function that Henry Kissinger had in mind when he supposedly told the Chinese leader Deng Xiaoping, "I think if we drink enough mao-tai we can solve anything."[68] Intoxication has therefore played a critical role in helping humans get past the cooperation dilemmas that pervade social life, especially in large-scale societies. For groups to move past suspicion and second-guessing, our sneaky conscious mind needs to be at least temporarily paralyzed, and a healthy dose of chemical intoxicant is the quickest, most effective, and most pleasant way to accomplish this goal.

PUKING AND BONDING

Sociality revolves around trust. It is therefore not surprising that liquid truth serums have always served as a powerful symbol of social cooperation and harmony. In ancient Mesopotamia, the distinctive shape of the beer vat served as a symbol for social interactions in general.[69] Early Chinese ritual gatherings, whether aimed at harmony between humans or between the living and their ancestors, were organized around alcohol, with ritual paraphernalia dominated by elaborate bronze drinking vessels. The joyful declaration "the spirits are drunk!" in an ancient ode celebrates the good favor of the ancestors and the establishment of harmony between the living and the dead. Banquets and drinking sessions have, across the world and throughout history, brought strangers together, united feuding clans, snuffed out disputes, and lubricated the creation of new social ties. Our modern word "bridal," for instance, comes from the Old English *bryd ealu* or "bride ale," which was exchanged between bride and groom to seal their marriage, and crucially the new bond between their families.[70]

The anthropologist Dwight Heath, a pioneer of the study of the social function of alcohol, notes that it has always played a crucial bonding function in situations where otherwise isolated individuals are required to get along—sailors in port, loggers just having come out of the woods, cowboys gathering at a saloon.[71] The Industrial Workers of the World (IWW) was a union in the early twentieth century that needed to solve a serious public goods problem: getting ethnically diverse, mutually suspicious workers with different trades and backgrounds to put aside their narrow personal interests and present a unified front in high-stakes collective bargaining against capital owners. The degree to which they relied upon heavy drinking, combined with music and singing, is reflected in the nickname by which they are best known today, the Wobblies, most likely a reference to their manner of stumbling from saloon to saloon.[72] These drunken, singing Wobblies, with their motto "an injury to one is an injury to all," were quite successful in bringing together up to 150,000 workers across a wide variety of industries and winning important concessions from employers.

In many cultures, epic drinking binges also serve a military purpose. Among medieval Celtic, Anglo-Saxon, and Germanic tribes, periodic blowout drinking sessions helped bind warriors to their lords and to one another, with the exchange of alcohol serving as a potent symbol of loyalty and commitment.[73] We noted above that George Washington, despite having defeated a Hessian army by taking advantage of their drunken state, saw alcohol as such an important component of military esprit de corps that he urged Congress to establish public distilleries to keep the fledgling U.S. Army well stocked with booze. Frederick the Great of Prussia, in 1777, issued a diatribe against the novel, and in his view dangerous, habit of drinking coffee instead of beer:

> It is disgusting to notice the increase in the quantity of coffee used by my subjects, and the amount of money that goes out of the country as a consequence. Everybody is using coffee; this must be prevented. My people must drink beer. His Majesty was brought up on beer, and so were both his ancestors and officers. Many battles have been fought and won by soldiers nourished on beer, and the King does not believe that coffee-drinking soldiers can be relied upon to endure hardships in case of another war.[74]

Other chemical intoxicants have also been used to create the particularly intense form of social bonding required for warriors. An early Spanish missionary to the New World noted that some indigenous groups used peyote before heading out to war. "It spurs them to fight with no thought of fear, thirst, or hunger," he reported. "And they say that it protects them from all danger."[75] The battle rage of the legendary "beserkers" of Norse legend was likely driven by psychedelics,[76] and the feared assassins of ancient Persia derived their name (Persian *hashashiyan*, Arabic *hashīshiyyīn*) from the intoxicant from which they drew their fighting spirit, hashish.

A common cross-cultural pattern is that drinking is more associated with men than women. In cultures where both men and women drink, men tend to drink considerably more. Physiological factors almost certainly play a role here.[77] Men are, on average, physically larger and therefore require more alcohol than women to get the same psychological effect. Probably a bigger factor, though, is that, in traditional patriarchal societies, it is men who are the primary actors in public and political life, who most have to navigate cooperation dilemmas with potentially hostile strangers.[78] In contemporary indigenous societies in the Andes, for instance, the anthropologist Justin Jennings writes, "Men are associated more with the consumption of alcohol than are women…While both sexes drink, a man's relationship with other men is affirmed through drinking. His ability to hold his liquor marks him as a man, and, through alcohol, 'friendships and agreements are sealed and kinship is acknowledged.'"[79] Classic anthropological work by Dwight Heath,[80] on the remote Camba people of the Bolivian Amazon, documented the manner in which Camba men use alcoholic binges, often drinking to the point of unconsciousness, to enhance their social solidarity and overcome interpersonal conflicts. Those who puke together stay together.

This is why strangers are typically welcomed with copious quantities of booze. Successfully making it through a night of heavy drinking is perhaps the quickest way to become accepted in a novel social environment. The anthropologist William Madsen, doing fieldwork in rural Mexico, drew an angry crowd after being spotted taking photos of a local religious ceremony. Backed against a wall at machete point by a group of men drunk on *pulque,* a traditional beer made from agave juice, he was spared only because an elder from a neighboring village, where he had been staying, announced, "Release our friend. He is not a stranger. He has drunk our *pulque.*" The machetes immediately disappeared, and everyone sat down to drink *pulque* together.[81] Sharing drink expands the circle of belonging and trust. It is revealing that perhaps our most ancient extant legal document, the Code of Hammurabi, specifically tasks tavern keepers with reporting conspiracies hatched over a few pints of their beer, upon pain of death.[82] The deep bonding power of alcohol is precisely the sort of tool useful to aspiring rebels or revolutionaries.

A failure to share alcohol, or to deny a proffered cup, is therefore a serious act of rejection or hostility. It can even bring divine retribution. Jennings reports an early seventeenth-century myth about a Peruvian deity who came to test the virtue of a community by appearing at one of their banquets as a poor, starving stranger. Only one man noticed him and welcomed him with the offer of a drink. When the god finally revealed himself and took out his wrath on the selfish banqueters, only this man was spared.[83] Similarly, a refusal to accept an offered drink is often seen as a grave insult. In early modern Germany, for instance, "Refusing a glass offered in fellowship was an affront to honor that could move men at all levels of German society to draw their swords, sometimes with fatal results."[84] Similarly dire consequences could result from declining the offer of a shot in an American frontier saloon.

Because of the intensity and perceived authenticity of trust and bonds forged with alcohol, repudiating a pledge sealed with wine or beer is an unusually powerful sort of betrayal. The archaeologist Piotr Michalowski reports a dramatically unpleasant example from ancient Sumer, recounted in a letter of complaint to a king who continues to maintain ties with a person named Akin-Amar:

"Is not the man Akin-Amar my enemy and is he not His Majesty's enemy? Why is he still in favor with His Majesty? One time that man stayed with His Majesty where he drank from the cup and raised it (in salute). His Majesty counted him among his own men, dressed him and gave him a [ceremonial] headdress. But he went back on his word and he defecated into the cup from which he had drunk; he is hostile to His Majesty!" A powerful image, indeed. One cannot imagine a stronger insult than the symbolic reversal of consumption by means of defecation. This is but metaphorical shorthand for the destruction of a whole symbolic system that was established through complex greeting ceremonies and semi-ritualized gift exchange.[85]

That is certainly one way to reverse a toast. Akin-Amar might have soiled his fancy headdress to send the same message, but it is the targeting of the bond created through communal drinking that really drives the message home.

In many, if not most, societies, alcoholic inebriation not only serves to create bonds between potentially hostile people, but is also viewed as a collective rite of passage, a test of one's character. The ability to hold one's alcohol is a sign of more general reliability or even moral virtue. One of my favorite lines about Confucius, found near the end of a long account of how fastidious he was about everything he ate and drank, is "only with regard to wine did he set no limits."[86] The fact that

Confucius could drink to his heart's content but never become unruly is a sign of his sagehood. Socrates was similarly praised for his ability to keep his wits about him despite partaking, as any proper Athenian man did, in marathon drinking events. "He will drink any quantity that he is bid," wrote Plato, "and never be drunk all the same."[87] Indeed, for the Greeks, the symposium, an evening of wine drinking led by a "symposiarch" who controlled the pace of drinking, served as a means for "testing people—a touchstone of the soul that is inexpensive and also harmless compared to testing people in situations where moral failure could result in serious harm."[88]

The sinologist Sarah Mattice observes that both ancient China and ancient Greece combined a demand that adults (at least adult males) get drunk together with an expectation that this would be an opportunity for them to display self-restraint and virtue under challenging conditions. In ancient China, "It could often be seen as an insult not to get drunk, but on the other hand, one was also not supposed to get sloppy, as it were, because that would impact the maintenance of deferential relationships." As for the Greek symposium:

> With a sober symposiarch in the lead—monitoring the character of those involved—citizens are given the opportunity to test themselves against the desire to succumb to pleasures, at the very point when their self-control is at its lowest. By drinking wine and being present in a situation where shamelessness has a tendency to reign, the citizens can develop a resistance to immoderate behavior, and so develop their character. In addition, because...symposia would be civic events, they also provide the opportunity for the citizens' virtue to be observed and tested.[89]

If participating in communal drinking sessions impairs your ability to lie, increases your sense of connectedness to others, and provides a test of your underlying character, it is then understandable why non-drinkers tend to be viewed with suspicion. "Water drinker" was used as a term of abuse in ancient Greece. Since ancient times, declining to participate in the frequent ritual toasts that punctuate a traditional Chinese banquet has been an act of almost unthinkable rudeness, one that would immediately propel you outside the pale of civilized society. This link between drink and fellowship remains a powerful one today in cultures across the world. The anthropologist Gerald Mars, in a study of the social dynamics of a group of longshoremen in Newfoundland, notes, "When early in fieldwork I asked a group of longshoremen why someone who was married, young, fit and hardworking—all well regarded qualities in a workmate—was nonetheless an outside man, the answer given was that he was a 'loner.' When I queried what form this took I was told, 'He doesn't drink—that's what I mean by a loner.' "[90]

We find a similar pattern in cultures where some other intoxicant has taken the functional place of alcohol. In Fiji, John Shaver and Richard Sosis observed that men who drink the most kava gain prestige in the community, and frequent drinkers cooperate most closely in collective horticultural work. Men who display *kanikani,* an unpleasant skin condition resulting from excessive kava consumption, are viewed with respect and seen as true "men of the village," trusted to support village values and conform to social expectations. The two anthropologists suggest that the social and reproductive benefits accruing to these men as a result of their kava-based prestige must outweigh the more obvious physiological costs, which are considerable.[91] Conversely, men who hold back in their drinking, or eschew kava ceremonies altogether, are viewed with suspicion and frozen out of many communal activities.

The social functions of intoxication are well summarized in the classicist Robin Osborne's musings on the ancient Greek symposium:

Intoxication was not something merely tolerated in others because of the pleasures it gave to the self. Intoxication also both revealed the true individual, and bonded the group. The intoxicated...faced up to how they ordered the world and where they belonged in that world; those who would fight, and die, together established their trust in each other by daring to let wine reveal who they were and what they valued.[92]

This is also the context in which to understand Ralph Waldo Emerson's comment on the role of the humble apple in early American society: "Man would be more solitary, less friended, less supported, if the land yielded only the useful maize and potato, [and] withheld this ornamental and social fruit."[93] Apple blossoms provided beauty, and the fruit cider and applejack. Beyond the obvious usefulness of staid maize and potatoes, then, Emerson discerned a more subtle function for beauty and intoxication, equally important as bread and potatoes for us social apes.

LIQUID ECSTASY AND THE HIVE MIND

Sobriety diminishes, discriminates, and says no; drunkenness expands, unites, and says yes. It is in fact the great exciter of the YES function in man. It brings its votary from the chill periphery of things to the radiant core. It makes him for the moment one with truth.

—William James[94]

"Ecstasy" comes from the Greek *ek-stasis,* or literally "standing outside oneself." Beyond allowing potentially hostile individuals to better trust and like one another, extreme levels of intoxication—especially when combined with music and dance—can be a tool for effectively erasing the distinctions between self and other. Surrendering to the abandonment that comes with drunkenness thus often serves as a cultural signal that one has become fully identified with, or absorbed into, the group. Writing of the use of *chicha* among traditional cultures in the Andes, Guy Duke observes:

> In the Andes, public drunkenness was a central aspect to religious and social life...Intoxication was seen as a means of gaining a deeper connection to the spiritual realm and no ritual took place without inducement of intoxication among the participants...The purpose was to get as drunk as possible and show one's inebriation publicly as a sign of immersing oneself in the ceremony...Not only was ritual public drunkenness sought, in many cases it was mandatory.[95]

In a sample of 488 small-scale societies for which relevant anthropological accounts exist, Erika Bourguignon found that 89 percent instituted ritual practices designed to produce states of dissociation or ecstatic trance, typically through group dance, song, and chemical intoxication.[96]

The widespread role of the first two, dance and song, to create ecstatic union has long been recognized in the anthropological literature.[97] "As the dancer loses himself in the dance," writes Alfred Radcliffe-Brown in a classic account of a culture in the Andaman Islands, "he reaches a state of elation in which he feels himself filled with an energy beyond his ordinary state...at the same time finding himself in complete and ecstatic harmony with all of the fellow members of his community."[98] Typically, the focus has been on the psychological and physiological effects created through rhythm, synchrony, and repetitiveness. The godfather of modern anthropology of religion, Émile Durkheim, viewed music, ritual, and dance as the key cultural technologies employed to create the "collective effervescence" that bound together traditional cultures. The influential theorist Roy Rappaport similarly argued that "the ritual generation of *communitas* often rests in considerable degree on ritually-imposed tempos, on their repetitiveness and, more fundamentally, on their rhythmicity."[99]

More recent work in the cognitive science of ritual has followed this lead, with researchers focusing on ritual components such as physical synchrony. One study, for instance, found that getting strangers to dance in sync with one another—as opposed to conditions where their dancing was partially or completely asynchronous—boosted their pain thresholds (a good proxy for endorphin activation) and reported feelings of social closeness. Other studies have found that tapping in synchrony with another person increases feelings of liking, interpersonal trust, willingness to help, and sense of similarity with another,[100] and that these feelings of prosociality might be quite expansive, extending beyond the in-sync partner to include others not involved in the activity.[101]

These are all important findings. However, it is surprisingly rare for theorists of religion or researchers in the cognitive science of ritual to acknowledge that in many (if not most) traditional rituals, the people dancing and singing and moving in synchrony are also *high as kites*. The ritual life of early Incan and Mayan cultures, for instance, centered on public rituals that employed dance and music to bring together the community and honor the gods. It also involved a degree of wild alcoholic inebriation that shocked early missionaries.[102]

In ancient Egypt, the Festival of Drunkenness was a major holiday commemorating the salvation of humankind when the fierce goddess Hathor was tricked into getting drunk on red-dyed beer instead of human blood. After some ceremonial preliminaries, it consisted primarily of everyone getting completely hammered in the Hall of Drunkenness, participating in ritually sanctioned sexual orgies, and then eventually falling asleep. As Mark Forsyth observes, "This was drinking with only one aim in mind: sacred drunkenness, and to be a holy drunk you have to be wholly drunk."[103] In the morning, an enormous image of the goddess was snuck into the hall, and then everyone was abruptly woken up, still drunk, by an ear-shattering din of drums and tambourines, to find the great goddess looming over them. This can't have been wholly pleasant, but it had to have been awe-inspiring. The goal, as Forsyth notes, is to, by so thoroughly shattering the ordinary, sober self, produce a moment of "perfect communion" with the goddess, and thereby with the community.

The efficiency with which alcohol and other substances decenter the self is the reason that intoxicant-fueled ecstasy is as ancient as human ritual itself. Jars containing our earliest documented alcoholic beverage—a "Neolithic grog," made of honey mead, rice beer, and fruit wine—from the Jiahu tomb (7000 to 6000 BCE) in the Yellow River Valley, were "carefully placed near the mouths of the deceased, perhaps for easier drinking in the hereafter," and the contents were no doubt also imbibed by those performing and attending the funeral.[104] The most dramatic archaeological remains from Bronze Age China are enormous, elaborate ritual vessels designed for serving and drinking alcohol. Late Neolithic and early Bronze Age tombs are packed with drinking paraphernalia, musical instruments, and food remains, suggesting that, from the beginning of documented Chinese history, the dead were sent off in wild bacchanalia that culminated with the attendees, drunk as skunks, tossing their cups into the grave.[105]

Some of the earliest evidence of grape wine production in the West, from a cave complex in Armenia dating back to 4000 BCE, suggests that the first domesticated grapes were being produced in elaborate winery-cum-mortuary facilities: The stomping troughs, fermentation vats, and drinking bowls were found next to an extensive graveyard, with drinking cups scattered in and among the graves themselves.[106] Indirect evidence of the ancient connection between chemical intoxication, ritual, and ecstasy is found in a remarkable potsherd, dating back to the early Neolithic (ninth millennium BCE), and found at a site in what is now modern-day Turkey not far from Göbekli Tepe. It shows two joyful individuals being accompanied in dance by a turtle, the presence of a dancing animal being interpreted by scholars as a sign of "altered states of consciousness" (Figure 4.1).

Figure 4.1. Potsherd from Nevali Çori, ninth millennium BCE (photograph by Dick Osseman, used with permission; https://pbase.com/dosseman)

Given what we know from other archaeological finds in the region, this ecstatic dance was likely fueled by massive quantities of beer.

And speaking of beer-induced hallucinations, we would be remiss in failing to note that plant-derived psychedelics of various sorts have also commonly been used as catalysts for ecstatic group experiences. When bringing together different, and possibly rival, groups in order to trade and find mates, small-scale Amazonian cultures have traditionally drawn upon *yajé*, a liquid containing one or more hallucinogens, to induce multiday collective trances, accompanied by songs and dance. "The principal objective of the total experience," notes Robert Fuller, "is to demonstrate the divine origin of the rules regulating social relationships."[107] Collective meaning is established through communal tripping.

A recent survey of the global ethnographic record focused on music found a strong relationship between mentions of musical performance and alcohol,[108] an indication that music-driven group synchrony and harmony are typically facilitated by healthy doses of booze. This study utilized a centralized database of ethnographic accounts from around the world, created and maintained by the Human Relations Area File (HRAF) project, where ethnographies are tagged with certain thematic keywords, such as "marriage" or "cannibalism." A survey of this database by one of my research assistants, Emily Pitek, also found that, of the 140 cultures where "ecstatic religious practices" are mentioned, 100 (71 percent) also note the presence of "alcoholic beverages," "drinking (social)," "drunkenness (prevalence)," "recreational and non-therapeutic drugs," and/or "hallucinogenic drugs."[109]

In some cultures, ecstatic group experiences are instead facilitated by pure stimulants. For instance, in the Fang culture of Gabon, a local stimulant, *eboka*, allows users to dance energetically all night, creating a state of "euphoric insomnia."[110] It must also be noted that there are cultural ecstatic practices that eschew chemical intoxicants altogether. Pentecostal services, for instance, involve intense and extended singing and dancing that can lead to speaking in tongues and other expressions of having been possessed by the Holy Spirit. As we discussed in Chapter Two, there is more than one way to skin a prefrontal cortex: Intense physical activity can have similar effects as one or two shots of whiskey.

However, as awkward dancers and tentative singers around the world and throughout history have always known, truly letting yourself go and feeling the music is a lot easier after a few shots of liquid courage. It is therefore surprising how little attention the anthropologic and scientific literature on ritual and group ecstasy gives to the role of mind-altering drugs. Perhaps this failure to adequately appreciate chemical intoxication reflects Puritanical discomfort with pleasure lurking in the background of scholarly discourse. In a modern scientific context, it is no doubt also driven by the practical recognition that it is hard to secure institutional approval to get human study participants hammered. (Of course, this impediment itself reflects the Puritanical attitude of review boards.) In any case, most current academic work on "collective effervescence" or the "hive mind" is radically incomplete, displaying a strange lack of interest in the nature and function of the liquid being sipped in between the singing and dancing. Nonetheless, it is simply a fact that cultures around the world have realized that the best synergistic effect is created when the independent psychological effects of rhythm and repetition are combined with powerful drugs.

The British anthropologist Robin Dunbar represents one exception to the otherwise typical neglect of intoxication. Dunbar and his colleagues see the physiological effects of alcohol, in particular, as a crucial component in social rituals. Specifically, they point to the endorphin release triggered by booze, especially when drinking is combined with music, dance, and ritual, as a crucial factor allowing humans to cooperate on a scale unattainable by our monkey or ape relatives. Endorphins and other opioids are stimulated naturally in most mammals by sexual intercourse, pregnancy, birth, and breast-feeding, and all play a strong role in both mate pair bonding and mother-infant bonding. What humans have figured out, however, is that a tasty liquid can be consumed to expand the reach of this "neurochemical glue."[111] One would expect that an increase in serotonin, another effect of alcohol and other intoxicants, would add to the bonding mix. In addition to enhancing individual mood, increased serotonin has been shown to reduce selfish behavior in Prisoner's Dilemma games, while depleting serotonin through blockers, such as tryptophan, has the opposite effect.[112] This synergy has perhaps found its perfected form in modern rave culture, where the powerful boost in serotonin created by MDMA intoxication is combined with driving, repetitive beats and group synchrony.[113]

Other intoxicants, like psychedelics, destabilize the sense of self even more powerfully than alcohol, completely blurring the distinction between self and other in a manner that bonds people to one another and promotes group identity.[114] The downside is that they really knock you on your butt for a long time. This is why psychedelic use tends to be reserved for either a specialized class, such as shamans, or confined to important but relatively rare ceremonies. The periodic ritual use of peyote in the Sierra Madre Occidental mountain range of Mexico, for instance, has ancient roots, and its social function is clearly aimed at so thoroughly breaking down the self that it can be merged into a harmonious whole. A mid-sixteenth-century account of the region describes one of these communal peyote rituals: "They assembled together somewhere in the desert, they sang all night, all day. And on the morrow, once more they assembled together. They wept; they wept exceedingly. They said [thus] their eyes were washed; thus they cleaned their eyes."[115] What is being washed away in the chemical rush of mescaline, the psychoactive component of peyote, are the sort of selfish desires and petty grievances that prevent apes with overgrown PFCs from surrendering to the group.

These ceremonies are still performed in these same mountain desert regions. At the culmination of one ceremony, described by the anthropologist Peter Furst, everyone present is required to publicly acknowledge any sexual transgressions they may have committed since their last confession. This is revealing: Sexual jealousy and conflict over mates are possibly the most powerful of centripetal forces that cause groups to splinter or fragment. As each person names their past lovers in front of their spouses or current partners, "no display of jealousy, hurt, resentment, or anger is permitted; more than that, no one is even allowed to entertain such feelings 'in one's heart.'" The participants emerge "purged" of transgressions, ritually purified in a manner reminiscent of Catholic confession.[116] This is an unusually effective way of heading off conflict before it can happen.

Psychedelics used in this sort of ritual context are so powerful a tool for disarming individual defenses and uniting the group that the peyote ceremony has spread to other American indigenous groups in need of a counterweight to the loss of cultural identity. As Robert Fuller observes, prior to 1890 the peyote ceremony was rare north of the Rio Grande. After 1890, as traditional tribal cultures came under increasing pressure—being stripped of their identities and herded into reservations—they turned to a form of the peyote ritual known as the Ghost Dance as a tool for forging a new group identity.[117] This "Americanized" peyotism remains a vibrant religious tradition in the Southwest, one that has had to assert its rights to ritual use of chemical intoxicants against a Puritanical federal government.

"Ecstasy!" wrote Gordon Wasson, an amateur mushroom enthusiast who is best known for championing the case that ancient Vedic soma was derived from the *Amanita muscaria,* or "fly agaric," toadstool. "In common parlance ecstasy is fun. But ecstasy is not fun. Your very soul is seized and shaken until it tingles. After all, who will choose to feel undiluted awe? The unknowing vulgar abuse the word; we must recapture its full and terrifying sense."[118]

True ecstasy is terrifying to the individual because it shatters the boundaries of the self. This is frightening and disorienting for an ape, but simply business as usual for a honeybee or ant. The ecstasy that comes with chemical intoxication causes not merely physical and mental pleasure, but effects a transformation that is crucial for achieving group cohesion. If we think of alcohol, for instance, as disabling negative barriers to cooperation (lying, suspicion, cheating), we have to also see its positive role in building affiliative, pair bond–like emotional ties between members of the group through the stimulation of endorphins and serotonin. Chemical ecstatic states are both a scalpel that disarms the self and the glue that binds suspicious, selfish apes into the cultural hive mind.

POLITICAL POWER AND SOCIAL SOLIDARITY

We've touched several times on what is possibly the oldest epic ritual site in the world, the stone enclosures and mysterious, monumental pillars erected at Göbekli Tepe (Figure 3.2).

Figure 3.2. Images from Göbekli Tepe (Klaus Schmidt/DAI; Irmgard Wagner/DAI)

Over 11,000 years old, Göbekli Tepe must have been created by hunter-gatherers, since it predates the advent of settled agri-culture. Its discovery a couple decades ago was therefore an important piece of evidence against the traditional view that certain key trappings of civilization—monumental architecture, elaborate ritual-based religion, and the brewing of alcohol—only could have developed once humans had attained the stability and access to resources brought about by the agricultural revolution. Beer before bread advocates see this site, with its stone basins that could hold up to forty gallons of liquid, scattered remnants of drinking vessels, and evidence of extensive feasting on wild animals, as an illustration of how ancient humans were first motivated to come together in large groups by the draw of intoxication and ritual, with agriculture coming after. It is revealing that there are no grain silos or other food storage facilities at Göbekli Tepe. "Production was not for storage," notes the archaeologist Oliver Dietrich and his colleagues, "but for immediate use."[119] In other words, people gathered in large numbers at this site for temporary, epic, blowout feasts, accompanied by dramatic rituals,[120] all of it likely fueled by generous quantities of booze.[121]

The booze served multiple functions. The appeal of drink and food drew together otherwise widely scattered hunter-gatherers from far and wide, creating the kind of workforce that could move, carve, and erect enormous sixteen-ton stone pillars. The monumental architecture, in turn, must have lent incredible authority and power to the organizers, while the intoxicant-fueled rituals conducted among these pillars created a sense of religious and ideological cohesion. Periodic alcohol-fueled feasts, after which the participants scattered again until the next ceremony, thus served as a kind of "glue" holding together the culture that created Göbekli Tepe and other sites in the so-called Golden Triangle where agriculture, and civilization, had its birth.

We see a similar connection between large-scale, centralized production of alcohol and the beginnings of political and ideological unity in other regions of the world where great civilizations independently arose. We have seen that the rulers of the Erlitou and Shang cultures in the Yellow River Valley of China appear to have derived their power from rituals powered by various beers and fruit wines, while the standardization and large-scale production of *chicha* was a crucial tool used by the Incas in the South American Andes to consolidate their empire. As Guy Duke writes:

Akha mama, the name of the fermenting agent used to begin batches of chicha, was also another name for Cuzco, the Inca capital... This highly symbolic alternate name shows the importance of chicha to the Inca on a number of different levels. On one level it shows the centrality placed on chicha by the Inca to the very process of ruling: without Cuzco, there would be no Inca and without chicha there would be no Cuzco. As well, through the implication that chicha originated, or was born, in Cuzco, the Inca Empire appropriated chicha and its social power as a means of establishing its legitimacy throughout the Andes: those who control the chicha control the Andes.[122]

The political function of alcohol is practical as well as symbolic. Whoever controlled the production of beer at Göbekli Tepe controlled the labor force that it attracted, and no doubt benefited practically from the ideological system created, reinforced, and spread by the resulting drunken, religious feasts. Incan emperors similarly used the promise of food and *chicha* to attract the massive number of laborers required to tend their cornfields and construct their own monumental architecture. The reliance of early Chinese rulers on group labor rewarded with alcohol is reflected in an ancient poem from the *Book of Odes* that reads,

In the south is the barbel
And, in the multitudes, they are taken into baskets
The host has wine
On which his admirable guests feast.[123]

To this day, throughout the world, large communal projects that rely upon unpaid labor, like constructing buildings or maintaining canals or irrigation channels, typically compensate the workers with massive, alcohol-heavy banquets or feasts funded by central authorities or local patrons.[124] Paul Doughty notes that the practice of "festive labor" in contemporary Peru, where volunteer work parties are treated to copious quantities of alcohol and music, remains the only way to accomplish large projects in the absence of a well-developed system of formal wage labor. In industrialized societies, where we have unions and 9-to-5 workdays with set wages and health care, drinking on the job is discouraged. In preindustrial societies, facilitating drinking on the job is the only way to get the job done.

The ability to provision large numbers of people with alcohol, as well as the food and entertainment that often come with it, is typically limited to local elites. Alcoholic feasting thus also serves as a way to announce, symbolize, and reinforce social status. This is especially the case because alcohol, unlike more mundane staples like bread or rice, is essentially a luxury good. Fermenting grains and fruit into beers and wines contributes to the concentration of wealth, in that fermented beverages magnify calories and vitamins while simultaneously compressing bulky, scattered biological resources into small, portable, and usually somewhat storable packages.[125] Even today, some degree of prestige and authority accrues to the provider of the keg at a university frat party. This is merely a pale, distant echo of the power enjoyed by the organizers of the feasts at Göbekli Tepe over 11,000 years ago, or the early Erlitou and Shang rulers in late Neolithic and early Bronze Age China, as they presided over epic, drunken rituals.

In parts of Iron Age Europe, it is worth noting that it was elite female rather than male tombs that included large and precious vessels for serving mead and wine. For instance, the lavish tomb of an elite woman buried in Vix, France, around 500 BCE, contained a chariot, gold jewelry, and other luxury items, including a wide variety of alcohol-related serving vessels imported from Greece. The most spectacular item, though, was an enormous *krater*, or bronze wine cask, over five feet tall and capable of holding 300 gallons of liquid, that was apparently produced in Corinth, Greece, around 600 BCE and transported to Vix in pieces. Such a massive and precious item, an exotic antiquity brought a long distance at enormous cost, must have served as a potent symbol of power, and probably featured prominently in formal ritual ceremonies. Michael Enright argues that burials such as these likely reflect a religious system where it was women who controlled access to alcohol, possibly in their role as priestesses charged with bonding together local males into cohesive war-bands.[126]

In traditional societies, the ceremonial use of alcohol also involves elaborate protocol and pageantry that serves to emphasize rank and social hierarchy.[127] From ancient Sumer to ancient China, drinking ceremonies were highly ritualized and carefully controlled by rulers and religious specialists, with an emphasis on highlighting status and rank. In most African societies, access to alcohol has traditionally been controlled by elites and often restricted to their use. There were exceptions for occasional public rituals, during which alcohol was shared with commoners, but still in a manner that advertised the differences in rank and the prestige of the rulers.[128] Elites, for instance, tend to drink first, and of the choicest beer or wine, and then have the doling out of booze to the masses directed by their representatives. Even in more egalitarian societies, the production of intoxicants such as palm wine brings status to the person who creates or purchases it. It is widely shared in communal celebrations, but in a way that allows the host to advertise his connections and gratitude toward those to whom he is offering the wine, and also in a manner that brings particular honor to the host and reinforces local status distinctions.[129] Other intoxicants, like kava, are used in a similar manner. In a traditional kava ceremony in Fiji, for example, men (and only men participate) are seated in a circular arrangement strictly dictated by rank, with everyone drinking in proper order and keeping their cup oriented toward the chief, who sits at the top, or "high," area of the ritual space.[130] By embedding intoxicant consumption in a ritual context, cultures not only bring people together, but also make it clear where they stand in the greater scheme of things.

CULTURAL GROUP SELECTION

Much of the anthropological literature on alcohol emphasizes the symbolic and political power that comes with control over its production and distribution, as well as the highly ritualized ways in which it is consumed. It is certainly important to recognize that alcohol is not merely a psychoactive chemical, but also serves as a bearer of cultural meaning. The source of the cultural meaning associated with intoxicants, however, is clearly grounded in their physiological effects. We should not fail to notice the conspicuous absence of kimchee- or yogurt-based super-cultures. The 300-gallon bronze vessel buried with the priestess of Vix was built to contain booze, not porridge; the fermenting vats and storage basins at Göbekli Tepe were not meant for sourdough bread. Ecstatic Shang elites tossed wine cups into the tombs of their deceased relatives, not millet bowls. Intoxicants have accrued enormous symbolic meaning because they *intoxicate* us. It is precisely the psycho-pharmacological efficacy of alcohol that allowed it to catalyze the rise of massive civilizations. In light of this, it's no surprise that cultures around the world quickly came to imbue alcohol—the great facilitator of civilization—with symbolic importance. Before civilization, intoxication.

This crucial functional role for alcohol and other intoxicants is slowly gaining wider acceptance in the anthropological community.[131] Proponents of the beer before bread hypotheses rightly emphasize how the increased cohesiveness and scale of intoxicant-using cultures would give them a distinct advantage in competition with other groups, allowing them to cooperate more effectively in work, food production, and warfare.[132] The inexorable pressure of cultural group selection would, in this way, encourage and disseminate the cultural use of intoxicants in the manner that we actually observe in the historical record, and that is completely inconsistent with any hijack or hangover theory of intoxication.

There is a very good reason we have historically gotten drunk. It is no accident that, in the brutal competition of cultural groups from which civilizations emerged, it is the drinkers, smokers, and trippers who emerged triumphant. In all of the ways outlined above, intoxicants—above all alcohol—appear to have been the chemical tool that allowed humans to escape the limits imposed by our ape nature and create social insect–like levels of cooperation. We have seen that traditional views about the functional benefits of alcohol consumption find confirmation in modern science. By enhancing creativity, dampening stress, facilitating social contact, enhancing trust and bonding, forging group identity, and reinforcing social roles and hierarchy, intoxicants have played a crucial role in allowing hunting and gathering humans to enter into the hive life of agricultural villages, towns, and cities. This process has gradually scaled up the scope of human cooperation, eventually creating modern civilization as we know it.

One might argue, however, that this is of merely historical interest. It is certainly possible that we no longer need intoxicants to continue doing all of this work for us. For instance, we now have at our disposal other ways to reduce stress and raise mood: Television and the internet, or antidepressants, may work just as well as a few pints of beer on that front. Perhaps they're even better. Modern banking systems and strong rule of law make us less reliant on handshakes or trusting the cut of someone's jib. Large public projects can now be funded by taxpayer dollars and completed by trained, sober professionals in return for salary and benefits. And perhaps, in a globalized world, the consolidation and monopolization of power by nationalist elites is something we want to move away from.

These are all reasonable observations. There are ways in which new technologies, less harmful and more targeted drugs, and modern institutions can provide many of the functional benefits historically provided by intoxicants, only without the nasty toxic part. I would argue, however, that we have not entirely outgrown our need for chemical ecstasy. Alcohol and other intoxicants can and should continue to play a role in our modern world. Indeed, in some ways we need them more than ever. There's a strong case to be made that chemical intoxication has not outlived its functional role, and there are plenty of reasons we should continue to get drunk.

CHAPTER FOUR

INTOXICATION IN THE MODERN WORLD

Stuck within the confines of the hijack or hangover theories of intoxication, it's hard to make the case that we should continue to get drunk or high. That hasn't stopped people from trying. Since pleasure per se is rarely accepted as a publicly defensible rationale, defenders of alcohol use, for instance, have generally focused on its supposed health benefits. Wine lowers cholesterol! It's good for your heart!

The scientific literature on the health benefits of alcohol consumption is, in fact, mixed and confusing. This is reflected in modern government policy, which has zigzagged between blanket condemnations of alcohol and recommendations for moderate consumption. In 1991, for instance, the U.S. federal government, in its official dietary guidelines, declared that alcoholic beverages have "no net health benefit," and that consumption of alcohol, in any amount, was not advised. By 1996 things had changed, with a U.S. federal health panel officially recognizing, for the first time, that moderate alcohol consumption might bring some health benefits.

Many are familiar with the so-called French paradox, which has been trumpeted with great fanfare by the wine industry. Despite the expected cardiac disaster that is traditional French cuisine, centered on butter, milk, and foie gras, the French have surprisingly low rates of heart disease. The claim was that at least part of the secret is the amount of wine, particularly red wine, drunk by the French, which appears to compensate for high levels of saturated fat. While the details of the French paradox have been disputed, research does suggest that moderate intake of any alcohol reduces the risk of coronary heart disease,[1] apparently by boosting the level of "good" HDLs. There is also some evidence for the long-term cognitive benefits of moderate alcohol use, including improved function on tasks such as memory or semantic fluency tests, as well as a decreased risk of depression.[2]

The happy notion of doctors prescribing two glasses of wine with dinner was shattered in 2018 with the publication of the study in the prestigious medical journal *The Lancet* mentioned in the Introduction. While acknowledging that moderate alcohol consumption might provide some modest health benefits, it noted that these were far outweighed by the massive costs in terms of accidents, liver damage, and other causes of early mortality. It concluded that the safest level of alcohol consumption was zero. In July of 2020, it was reported that the next update to the U.S. federal government's Dietary Guidelines for Americans will tack hard toward abstention, recommending that people limit themselves to one drink per day—"drink" being defined quite stingily as 12 ounces of beer, 5 ounces of wine, or 1.5 ounces of distilled liquor.[3] This is in keeping with a worldwide trend of government tightening up their recommendations for alcohol consumption.

The adoption of such guidelines typically inspires both cheers from anti-alcohol campaigners and much raging and gnashing of teeth among tipplers. The former include neo-Prohibitionists, such as the journalist Olga Khazan, who argues that alcohol should be treated on the same footing as other (mostly illegal) drugs. "Beyond how it tastes and feels, there's very little good to say about the health impacts of booze," she declares. "Alcohol is the one drug almost universally accepted at social gatherings that routinely kills people."[4] Among alcohol's defenders, some dispute the statistics employed in the *Lancet* study, or argue that authorities are underweighing the importance of coronary health or alcohol's beneficial effects on digestion.

For our purposes, the most important thing to note is that this whole kerfuffle serves as a perfect example of how a failure to consider the functional, social benefits of alcohol can seriously skew public debate on the topic. There is no need to quibble around the margins about HDL levels. The most important thing that neo-Prohibitionists and health authorities alike fail to consider in coming down on the side of total abstinence is that the obvious physiological and psychological costs of alcohol must be weighed against their venerable role as an aid to creativity, contentment, and social solidarity. Once we recognize the functional benefits of intoxication—its role in helping humans to adapt to our extreme ecological niche—the argument that we should strive for a completely dry world is difficult to sustain.

We saw in Chapter Three how alcohol and other chemical intoxicants catalyzed and supported the rise of civilization itself. Chapter Five will delve into the dangers that alcohol and alcohol-driven behavior present to both individuals and society, especially in a world awash in distilled liquors and bereft of traditional social controls. Here, though, I would like to make the case that alcohol and other intoxicants have not outlived their usefulness. The difficulties involved in being the creative, cultural, and communal ape have not disappeared simply because we now have access to TED talks, Zoom, and (at least in Canada) universal health care. It is still hard being human. This means that, despite the trouble he inevitably brings in his wake, we still need Dionysus to play a role in our lives.

WHISKEY ROOMS, SALOONS, AND THE BALLMER PEAK

We've seen that contemporary cognitive science and psychology suggest that the link between intoxication and creativity is no myth. People who have reduced cognitive control, including children, do better at lateral thinking tasks like the Remote Associates Task (RAT), where you must come up with a fourth word (e.g., PIT) that ties together an otherwise nonsensical collection of target words (PEACH, ARM, TAR). People who have had their cognitive control abilities temporarily impaired in some way—for instance, by having their PFC knocked offline by a transcranial magnet—also display enhanced performance. Given the relationship between decreased cognitive control and thinking outside the box, we should expect that, in small to moderate doses, chemical intoxication would increase individual creativity.

That is, indeed, what we find. The first study to directly measure the effect of alcohol on creative thinking, entitled "Uncorking the muse: Alcohol intoxication facilitates creative problem solving," was published in 2012 by Andrew Jarosz and colleagues.[5] Subjects were brought into the lab, weighed, given some bagels to buffer their stomachs, and then asked to complete a working memory capacity task, a commonly used measure of executive function. This provided a pre-intoxication benchmark of their cognitive control abilities. Then came the alcohol: a series of vodka and cranberry cocktails over a ten-minute period, during which subjects were distracted by watching the animated film *Ratatouille* (!). The alcohol dosage was calibrated to their weight in order to eventually bring them to between 0.07 and 0.08 percent blood alcohol content (BAC). This is pretty well lubricated but not completely drunk, the point at which most jurisdictions make it illegal to drive. While their buzz was coming on, they completed a series of distraction tasks. Once intoxication was expected to peak, subjects were given a second cognitive control task, had their BAC measured, and were given a series of RATs to complete. They also had to report on whether they figured out the solution to the RATs through careful, step-wise reasoning or simply in a flash of insight. A sober control group skipped the bagels and drinks, but still got the animated rat movie and performed the same tasks.

Since both drunk and sober participants completed a working memory task before the drinking began, drunk subjects could be compared to sober counterparts with similar levels of executive function. As expected, once the drinking group got drunk, they underperformed the controls on the second executive function task—the alcohol did its job of taking the PFC temporarily offline. The drunks, however, smoked the sober controls on the RAT task, solving more RATs and more quickly. They were also more likely to report having solved the tests in moments of inspiration, with the answers just popping into their heads.

This is only one study, with a fairly small number of participants. Coupled with a much larger body of work linking decreased cognitive control to enhanced lateral thinking, though, it bolsters ancient, cross-cultural views about alcohol and creativity. We can add to it one other alcohol study, "Lost in the Sauce"[6] (if nothing else, alcohol increases the creativity of study titles), which provides strong indirect evidence for the role of alcohol as muse. Michael Sayette and colleagues had subjects consume either an alcoholic or virgin cocktail and then perform a task where they had to read selections from *War and Peace* and press a button labeled "ZO" whenever they caught themselves "zoning out," or mind wandering. Additionally, every few minutes their mind wandering was probed by the experimenters: a tone would go off and the question, "Were you zoning out?" would pop up on the screen. The mildly drunk subjects drifted off more often, and they were also less likely to notice their minds wandering. As we discussed earlier, there is a solid body of evidence that mind wandering and creativity go hand in hand, so enhancing the first should boost the second. A mind unmoored from the task at hand and roaming free, not even aware that it is wandering, is a mind ready to produce creative insight.

After I presented these studies in a talk at a Google campus some years ago, I was immediately brought by my excited hosts to their impressive whiskey room. This is where coders retire for a dram of liquid inspiration when they run into a creative wall. I was also introduced on this visit to the concept of the Ballmer Peak (Figure 4.1). Attributed, perhaps apocryphally, to Steve Ballmer, the former Microsoft CEO, this is the extremely high but very narrow peak of a curve describing one's level of coding skill as a function of blood alcohol content (BAC):

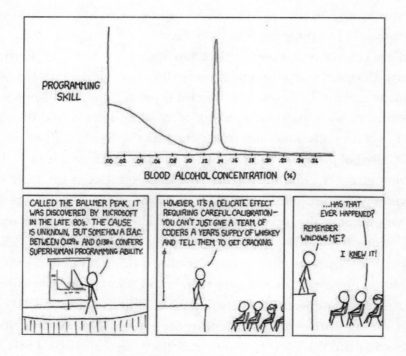

Figure 4.1. xkcd on the Ballmer Peak (xkcd.com)

Legends circulate of coders, in proper engineering fashion, hooking themselves up to alcoholic IV drips to remain hovering around the BAC sweet spot. Although Randall Munroe, the creator of the xkcd cartoon, jokes about the inadvisability of throwing a team of coders in a room filled with whiskey, this is, in fact, precisely what Google does on a regular basis. This leads to our second point about alcohol and creativity: The boost that alcohol provides to individual creativity is magnified and enhanced when people get drunk in groups. The Google whiskey room is, crucially, a *communal* space, filled with informal group seating arrangements—not a place to get drunk alone. The coders I spoke to said that they used it primarily in teams, as a place to get away from their screens, relax their minds, and allow new solutions to come to them when facing seemingly intractable problems. Spaces that allow for both communication and easy access to alcohol can act as powerful incubators for collective creativity.

A whiskey room with beanbag chairs and foosball tables is one such space. So is basically any traditional salon, banquet table, pub, or bar. The layout of your typical modern drinking establishment, with seating configured to allow people to comfortably eat and drink in groups of various sizes, is perfectly designed to catalyze group creativity. In downregulating our cognitive control and boosting our mood and energy level, alcohol not only opens the mind to creative insight but also lowers the barrier to communicating these insights to others. "Stupid" ideas seem less stupid after the second drink; by the third drink, your senior colleague seems less intimidating. Alcohol also increases general risk-taking. In Chapter Five we will discuss how this can have significant negative consequences when it comes to sex and operating heavy machinery, but in the world of ideas embracing risk is a positive thing. In any case, there is considerable anecdotal evidence that, by freeing up individual minds, lubricating the flow of ideas from person to person, and reducing self-consciousness and inhibition, communal alcohol consumption is a key driver of cultural innovation.

Iain Gately notes that in ancient Persia no important decision was made without being discussed over alcohol, although it would not actually be implemented until reviewed sober the next day. Conversely, no sober decision would be put into practice until it could be considered, by the group, while drunk.[7] From ancient China and ancient Greece to modern-day Silicon Valley, communal thinking and group drinking have always gone hand in hand. Gately has also argued that more recently the "water trade," the infamously alcohol-soaked, after-hours but mandatory drinking sessions endured by Japanese salarymen (and they were almost all men), was a key driver of Japanese industrial innovation in the 1970s and '80s. One of its functions was to suspend social hierarchy norms in order to allow innovative ideas to flow from junior to senior employees. "Alcohol was the lubricant that enabled the Japanese business machine to run smoothly," Gately notes. "While the gerontocracy demanded and received respect for their years when behind their desks, after work, off premises, and over a drink, they let their young Turks speak."[8] The anthropologist Philip Lalander documents a similar dynamic among young Swedish bureaucrats in the 1990s, where periodic heavy drinking sessions after work allowed them to escape workplace norms, mock the social hierarchy, and give voice to subversive opinions or suppressed desires—all in a safe context, with everyone equally defenseless, PFCs drowned in vodka.[9]

People say a lot of stupid things when drunk. But novel or innovative ones tend to rise to the surface of the torrent of ideas that flows back and forth in a group when everyone is relaxed and happy, defenses down and open to insight. One of the most significant political ideologies of modern times, communism, was forged by Friedrich Engels and Karl Marx over "ten beer-soaked days"[10] in Paris in 1844; it's no surprise that the salons of Paris, well provisioned with alcohol, were hotbeds of intellectual and artistic innovation. The saloon played a similar function in pre-Prohibition America. As the novelist Jack London wrote, "Always when men came together to exchange ideas, to laugh and boast and dare, to relax, to forget the dull toil of tiresome nights and days, always they came together over alcohol. The saloon was the place of congregation. Men gathered to it as primitive men gathered about the fire."[11] And we should never forget that our word "symposium," the quintessential forum for academic exchange and intellectual innovation, originally referred to the apex of ancient Greek sociality, the wine drinking party.

It is hard to find systematic data on alcohol and cultural innovation, but a recent working paper by the economist Michael Andrews, "Bar Talk: Informal Social Interactions, Alcohol Prohibition, and Invention," represents one fascinating attempt to do so. Andrews begins by reviewing the literature in economics on "collective invention,"[12] which documents the manner in which informal, chance social interactions drive innovation and growth. This is one fairly obvious reason that dense urban areas, particularly if they include a mix of industries and academic institutions, tend to be sources of new ideas.[13] Andrews cites an observation made in 1890 by the venerable economist Alfred Marshall, who noted that when people and companies are brought together in dense urban centers...

[t]he mysteries of trade become no mysteries; but are as it were in the air, and children learn many of them unconsciously. Good work is rightly appreciated, inventions and improvements in machinery, in processes and the general organization of the business have their merits promptly discussed: if one man starts a new idea, it is taken up by others and combined with suggestions of their own; and thus it becomes the source of further new ideas.[14]

"But how do ideas get into the air and spread from person to person?" Andrews asks in his paper. "Nineteenth century machine tool manufacturer and inventor Richard Roberts suggests that, rather than moving through the air, ideas are transmitted through the tap: 'No trade can be kept secret long; a quart of ale will do wonders in that way.'"[15] Andrews notes that there is a growing literature on the importance of bars as gathering places for creative individuals, and a wealth of examples of inventions or new technologies being hatched in saloons, (alcohol-serving) cafés, or bars.[16]

Not content with this anecdotal evidence, Andrews decided to turn to a useful natural experiment to actually test the idea that the communal consumption of alcohol is a driver of innovation: American Prohibition. Although we now tend to think of Prohibition as a single event, imposed by the U.S. federal government in 1920, the Prohibition movement had a much longer history in the States, with local, county-level bans on alcohol production and consumption going as far back as the early 1800s. It's worth noting that pre-Prohibition booze stockpiling, home distilling, and an active black market meant that legal prohibition never really eliminated drinking. Wherever it was enacted, however, it very effectively put an end to *social* drinking by killing off the saloon and forcing drinkers into the isolation of their own homes or small, private gatherings.

Andrews took advantage of the variation when U.S. counties adopted alcohol restrictions to look at the effect of prohibition on an excellent proxy of innovation, the rate of new patent registration, for which he could also obtain data on a county level. Using the state-level imposition of prohibition as a starting point, he compared counties that had been consistently dry for a long period of time to counties that had been "wet," but were now suddenly forced to close their saloons and other public drinking venues. He found that prohibition reduced the number of new patents by 15 percent annually in previously wet counties relative to previously dry counties. After three years of prohibition, however, the gap gradually closed. Andrews speculates that this eventual uptick in innovation might be driven by the gradual emergence of speakeasies and other illegal social drinking venues that sprang up to replace saloons.

I am a professor, not an inventor, artist, or businessperson, but a strong case can be made that alcohol-facilitated sociality plays a similarly pivotal role in academic innovation. My graduate seminars in the 1990s often ended with all of us, students and faculty, decamped to the campus pub, where the debate started in the seminar room continued over pitchers of beer and bar snacks— and often, after a pint or two, went off in unexpected and creative directions. On one such occasion, I witnessed firsthand a powerful demonstration of the role of the modern-day salon or saloon in driving innovation. Several years after I started at the University of British Columbia, which had been bizarrely devoid of informal drinking establishments, a large, comfortable pub finally opened. It was perfectly situated next to the bus loop, an ideal place to gather at the end of the day before heading home, and a small number of us were inspired to institute a weekly get-together on Friday afternoons. We were an intellectually diverse group, and there was no agenda, other than to have a few drinks and appetizers and chat. Over the next two years, however, the ideas and collaborative ties generated in those pub chats led to the establishment of a new center at UBC, a $3 million federal grant, an award-winning journal article, a slew of high-impact studies, and a massive new database project. There is simply no way these conversations would have been catalyzed by a new Starbucks or bubble-tea establishment—we needed a pub. This is precisely why, in Oxford colleges, evenings of discussion and debate formally begin with the Latin declaration, *nunc est bidendum* ("Now is the time for drinking").[17]

We'll discuss in Chapter Five the dark side of alcohol-based sociality. In recent years, a long-overdue discussion about the negative and sometimes criminal and tragic outcomes of mixing work and booze have come to the forefront of public consciousness. This is certainly a good thing. After a broad and careful consideration of the relevant pros and cons, we may very well conclude that graduate students should never mix in the pub with their professors, or that professional events should be completely dry. But any such discussion needs to take into account not only the more obvious costs of mixing work and drink, but also the subtle benefits we'd be losing. With gently downregulated PFCs, students speak out more freely, make intellectual connections with one another, and get to witness their mentors working things out on the fly, partially and temporarily free from the fetters of academic hierarchies. Colleagues float ideas that would otherwise never bubble up into consciousness and recklessly venture out of their intellectual safe zones, blundering across disciplinary boundaries that often desperately need to be crossed.

This book was written in the midst of the Covid-19 pandemic. It will take years to understand the various ways in which this crisis may have negatively impacted innovation. More obvious and dramatic factors, like the stress of caring for sick loved ones or homeschooling children, clearly slash productivity and narrow one's focus. Less obvious, perhaps, is the way in which the widespread and abrupt transition from in-person meetings to Zoom and Google Hangouts has changed the way that people talk and think. Wide-ranging chats over a few beers, sprawling over an hour or two, have been replaced by shorter video meetings focused on a specific set of agenda points. In this artificial medium, people have trouble naturally interrupting one another or smoothly navigating shifts in topic or speaker. This is one of the ways in which the Covid-19 crisis, like American Prohibition, might provide an excellent natural experiment demonstrating how meeting in person, often over alcohol, enhances both individual and group creativity.

TRUTH IS THE COLOR BLUE:
MODERN SHAMANS AND MICRODOSING

Alcohol has dominated our story for good reason. It is far and away the most widespread, popular, user-friendly, flexible, and multi-purpose intoxicant known to humankind. Looking toward the future, however, it is worth focusing exclusively for a moment on hallucinogens or psychedelics. As traditionally used, they are very difficult to integrate into everyday life. The incredible disconnection from reality that they produce not only limits their social usefulness, but it also makes one dubious about the validity of the insights they produce. However, modern tweaks in the way psychedelics can be consumed might make them both more user-friendly and more useful.

I will confess to having dabbled in psychedelics as a young adult, usually in certain places of unusual natural beauty near my home in the San Francisco Bay Area, particularly the west face of Mount Tamalpais and Limantour Beach in Point Reyes National Seashore. I would often take a notebook with me to record my thoughts and "insights." During one particularly epic experience, I was seized by the conviction that the answer to all of life's questions, the key to understanding all reality, was the realization that truth was the color blue. Over twenty notebook pages I carefully and definitively proved this assertion, supported by several diagrams and some mathematical equations. I remember thinking that my university graduate program would, upon the publication of this treatise, be compelled to not only immediately grant me my Ph.D., but also make me a full professor.

As one would expect from the central argument, the essay seemed rather less world-shaking the next morning. It was only years later that I ran across an account by the great psychological pioneer, William James, of a nitrous oxide trip he had experienced. He was similarly convinced during the experience that he had discovered the secret of the universe, but upon reading his notes the next morning found only this:

Higamus, hogamus,
Woman's monogamous,
Hagamus, higamus,
Man is polygamous.[18]

Indeed. James's limerick fell somewhat short of tearing back the curtains of everyday life to reveal the contours of true reality. *Truth Is the Color Blue* likewise did not, as it turns out, win me instant fame and professional success, and languishes unpublished to this day.

What I *did* bring back from these trips, however, were new and important insights about my personal life, my past, and where I was heading in the future. These coherent insights typically only crystallized slowly, at the very end of the trip or even the day after, like pieces of shattered glass settling into a recognizable mosaic pattern. James similarly re-evaluated the long-term value of psychedelic trips, noting that "they may determine attitudes though they cannot furnish formulas and open a region though they fail to give a map. At any rate they forbid a premature closing of our accounts of reality."[19] Aldous Huxley, reflecting on an unusual flower arrangement seen during a mescaline trip, felt that he had been given the glimpse of "what Adam had seen on the morning of his creation—the miracle, moment by moment, of naked existence."

> To be shaken out of the ruts of ordinary perception, to be shown for a few timeless hours the outer and the inner world, not as they appear to an animal obsessed with survival or to a human being obsessed with words and notions, but as they are apprehended, directly and unconditionally, by Mind at Large—this is an experience of inestimable value to everyone and especially to the intellectual.[20]

It is this idea of being radically "shaken out of the ruts of ordinary perception," to be open to new universes of thought, that is the particular gift of powerful psychedelics. We have noted that they have been used for millennia by shamans to bring back answers and insights from the spirit realm. We might see the "spirit world," in a modern context, as the wildly diverse, fragmented, technicolor, non-linear landscape of the human brain radically freed of cognitive control. Psychedelics are powerful depatterning or entropy-inducing agents, seriously disrupting the normally orderly flow of neural traffic directed by the PFC. The removal of the playground monitor allows confused and promiscuous cross-talk between brain regions that normally communicate through carefully regulated channels.[21] The result is mostly spatial disorientation, wildly confused perception, and conceptual nonsense. But the thoroughly shaken brain sometimes finds itself resettling into a usefully different configuration.

When attempting to explain the rise of Silicon Valley, where ideas and inventions that have thoroughly transformed the modern world were hatched, pundits typically cite the presence of Stanford University, or the mild climate capable of attracting bright people from all over the world. Less commonly mentioned, but probably no less important, is its proximity to America's psychedelic epicenter, San Francisco. As the writers John Markoff and Michael Pollan have documented, psychedelics—primarily pharmaceutical-grade LSD provided by a mysterious, colorful figure named Al Hubbard—played a central role from the very beginning of Silicon Valley's rise.[22] Ampex, an innovative, but now mostly forgotten, Silicon Valley–based manufacturer of storage devices, has been dubbed the "world's first psychedelic corporation" because of the weekly workshops and retreats it organized around LSD use in the 1960s. LSD was instrumental in the creative design process that gave rise to circuit chips, and Apple founder Steve Jobs claimed that his experiments with LSD ranked as some of his most important life experiences.[23] The synergy between San Francisco drug-based hippy culture and Silicon Valley innovation has been replayed in other places around the globe, from Berlin to Beijing, where intoxicant-heavy underground or bohemian cultures have rubbed shoulders with new industries dependent on creative insight rather than manufacturing muscle.[24]

A modern twist in hallucinogen use—a trend pioneered, as one might expect, in Silicon Valley—is making psychedelics easier to integrate into everyday life through the practice of "microdosing."[25] Microdosing involves taking frequent but small amounts of purified LSD or psilocybin, on the order of one-tenth of the normal dose, to induce mild, but sustainable, highs. The journalist Emma Hogan has documented widespread microdosing among knowledge workers in the San Francisco Bay Area.[26] One interviewee, "Nathan," credits microdosed LSD with increasing his productivity, giving him a creative edge, and magnifying his impact at investor-pitch meetings. "I view it as my little treat. My secret vitamin," he told her. "It's like taking spinach and you're Popeye." Hogan quotes an observation by Tim Ferriss, an angel investor and

174

author, that "the billionaires I know, almost without exception, use hallucinogens on a regular basis."[27]

These anecdotal accounts of microdosing enhancing creativity are supported by preliminary survey evidence. One recent study of online respondents[28] contrasted the performance of self-reported microdosers and individuals who had never microdosed on the Unusual Uses Task (UUT). It found that microdosers generated responses that were rated as significantly more uncommon, unexpected, and clever than those of their non-microdosing peers. Another study of Dutch microdosers in a naturalistic setting[29] found that administering a microdose of psychedelic mushrooms improved performance on two measures of creativity. The researchers in both studies acknowledged their respective limitations: The first relied on correlation and self-reports from volunteer online subjects, and the second lacked a non-microdosing or placebo control. Both, however, take us beyond mere anecdotal evidence. We'll learn more from randomized, placebo-controlled laboratory studies, several of which are currently planned or underway.

Modern science has given us the ability to purify the active ingredients in traditional psychedelics and deliver them in precise doses. If, eventually, this makes these drugs more amenable to everyday use, they might offer considerable advantages over alcohol. Despite lurid reports in the 1960s about LSD-induced insanity or tripping teenagers leaping off roofs, psychedelics are considerably safer, in most regards, than alcohol or cannabis. They are non-addictive, selectively target certain parts of the brain rather than playing havoc with the entire brain-body system, and cause no known side effects. In a 2009 briefing[30] the U.K.'s top drug adviser, Dr. David Nutt, ranked LSD (along with cannabis and MDMA) as less dangerous than alcohol and tobacco, although he was later forced to resign because of the resulting controversy.

As the creative and cultural species, we need all of the

innovation we can get. The neural reshuffling that psychedelics induce in individual brains could have important effects on group creativity. As Michael Pollan puts it, "Entropy in the brain is a bit like variation in evolution: It supplies the diversity of raw materials on which selection can then operate to solve problems and bring novelty into the world."[31] Pollan's popular account of psychedelics was inspired in part by the work of Giorgio Samorini, who has similarly argued that chemical intoxicants have played a crucial role, especially in times of rapid change, as a "depatterning factor" that increases cognitive and behavioral diversity in many animal populations, including humans.[32]

In his typically more colorful fashion, the 1960s LSD guru Timothy Leary once declared:

> "Turn on" means to contact the ancient energies and wisdoms that are built into your nervous system. They provide unspeakable pleasure and revelation. "Tune in" means to harness and communicate these new perspectives in a harmonious dance with the external world. "Drop out" means to detach yourself from the tribal game...In every generation of human history, thoughtful men [*sic*] have turned on and dropped out of the tribal game, and thus stimulated the larger society to lurch ahead.[33]

"Tribal game" is an excellent shorthand for the imperatives of our narrowly goal-focused and selfish primate nature. Leary's infamous motto, "Turn on, tune in, drop out," can thus be seen as an encouragement to be less like chimps and tap, instead, into our capacity to be creative, cultural, and communal primates. Sexism and woo-woo New Age language about "ancient energies and wisdoms" aside, we couldn't have a better expression of the role that powerful brain depatterners have historically played in accelerating the pace of cultural evolution. Especially given the enhancement in efficacy and ease of use that has been made possible by modern science, they should continue to play a similar role in today's world.

WHY SKYPE DIDN'T ELIMINATE BUSINESS TRAVEL

In 1889 Jules Verne predicted that the "phonotelephote"—essentially a dedicated videoconferencing device that he imagined would become commonplace by the year 2889 (!)—would make business travel obsolete.[34] We didn't have to wait a thousand years. Videoconferencing became a real technology in 1968 with AT&T's "Picturephone." The advent of Skype and other videoconferencing technologies in the mid-2000s brought phonotelephotes into every home that had access to a decent internet connection. Each new advance in remote teleconferencing capacity is accompanied by renewed predictions of the demise of business travel. Yet the fact is that, at least until the global Covid-19 pandemic hit in 2020, business travel has done nothing but steadily increase. Given the expense, hassle, and physiological toll of traveling, especially between very different time zones, this is genuinely puzzling. Why fly from New York to Shanghai to meet a potential business partner when you could just call or Zoom?

The puzzle of business travel is fundamentally related, I would argue, to the puzzle of why we like to get drunk. Neither makes practical sense unless we discern the cooperation problems to which they are a response. For simple and low-stakes long-distance transactions, like purchasing a book or sweater online, I am happy to rely on whatever enforcement mechanisms are at the disposal of eBay or Amazon should my counterparty prove untrustworthy. On the other hand, if I am entering into a long-term, complex venture with a company in Shanghai, where the impact of screwups or corner-cutting or backstabbing or simple fraud is multiplied a thousandfold, I need to know that the people I'm dealing with are fundamentally trustworthy. Yes, we will both sign a contract. But the loosely woven lattice of even the most comprehensive explicit agreement still allows multiple degrees of freedom. For anything more complicated than a one-off purchase of buttons or zippers, I am going to want to know who, really, I am getting into bed with.

One of the most effective mechanisms human beings have invented for assessing the trustworthiness of a new potential cooperator is the long, drunken banquet. As we have seen, from ancient China to ancient Greece to Oceania, no negotiation was ever concluded, no treaty ever signed, without copious quantities of chemical intoxicants. In the modern world, with all of the remote communication technologies at our disposal, it should genuinely surprise us how often we need a good, old-fashioned, in-person drinking session before we feel comfortable about signing our name on the dotted line.

This is not a foolish desire: As we've seen, a PFC-impaired person is a more trustworthy one. We've discussed how depressing PFC function makes it harder for a person to lie. Perhaps surprisingly, it has the opposite effect on lie *detection*.[35] It's actually more difficult for us to accurately evaluate the truthfulness of a statement when we are focused exclusively on doing so. We do a better job detecting lies when we are distracted by other stimuli—for instance, trying to get the bartender's attention or savoring an appetizer—and then asked later if the person we were speaking with was being honest. Our unconscious selves are better lie detectors than "we" are, and they are at their best when our conscious mind is temporarily sent to its room. The ancient intuition that alcohol reveals the "true self" is also more than just folk wisdom. Reduction of cognitive control results in disinhibition, a state where dominant tendencies that might otherwise be reined in by the PFC are set loose. In the absence of strong situational cues, for instance, people become more aggressive when drunk only if they are predisposed to aggression in general.[36] You may seem like a nice person on the phone, but before I really trust that judgment I would be well advised to reevaluate you, in person, after a second glass of Chablis.

The usefulness of in-person intoxicant consumption is not limited to the business world. An informal intelligence-exchanging alliance between European spy agencies that began in the late 1970s, and is still going strong, came to be referred to as Maximator. This is the brand name of a local, powerful doppelbock served in the suburban Munich pub where the initiative gradually emerged over rounds of beer.[37] It is hard to imagine spies from Denmark and Germany and the Netherlands overcoming mutual suspicions and confiding classified intelligence over coffee and pastries. Presumably these spooks were drinking on the government dime, as European governments and institutions tend to view alcohol as an integral part of normal human sociality.

Forging the international academic collaboration begun at UBC that I mentioned above presented more challenges: When it comes to alcohol, Canadians are almost as Puritanical as Americans, and Canadian federal regulations forbid spending grant funds on alcohol. In my view, this presents a serious barrier to scientific progress.[38] Tasked with forging a research partnership across different countries and disciplines, my colleagues and I saw only one response: We had to pony up personal money for a dedicated alcohol fund. As someone who has spent the last decade or so running these sorts of large, international projects, I can say with some certainty that it is next to impossible to get potentially rival research groups to cooperate without in-person socializing, good food, and judicious doses of liquid neurotoxin.

Appreciating the positive functions of alcohol-mediated, in-person socializing should also inform the current heated debate about professional travel and its impact on carbon emissions. Some climate activists have recently begun pushing for an end to in-person conferences, arguing (correctly) that it is an enormous source of greenhouse gas emissions and (less convincingly) that in-person interactions can simply be replaced with virtual ones. It certainly is the case that, in an age where lightning-fast and relatively ubiquitous internet connections make teleconferencing a viable option for academics and other professionals, it might seem wasteful to fly thousands of miles to sit in a room to hear a lecture or have a meeting around a table. If conferencing were only about the exchange of abstract information, there does seem to be more efficient ways to make this happen. What is missing in these discussions is a proper appreciation of the full range of reasons people have for attending conferences.

Yes, the cutting-edge research conveyed in the formal talks is important, although the gist of many of these presentations is now also live-tweeted by other audience members. Perhaps just as important, however, is the socializing, and facilitating this is not merely a frivolous waste of taxpayer money or unmitigated tragedy for the planet. A unique intellectual benefit provided by in-person academic or professional conferences is the networking, brainstorming, and idea-honing that goes on over meals, over coffee breaks, and most of all at informal venues, as the day draws to an end and the intoxicants come out. Innovation is the lifeblood of modern economies, and at the center of academic progress. Teleconferencing is rational, cheap, and environmentally friendly— not to mention lifesaving during the Covid-19 crisis in the midst of which I am writing—but even an extended Zoom session won't involve the participants getting buzzed and sketching out a new study design on a cocktail napkin.

Videoconference happy hours have become popular as an

attempt to recapture the energy and dynamics of the real thing. However, physically isolated individuals mixing their own single-serve martinis, on different continents and in various time zones, have to endure video glitches, bad audio, and, even with the best internet connection, a subtle but corrosive delay in response time that makes it difficult to properly time interruptions or changes in topic. Even the best videoconference is a poor substitute for the visceral buzz of interpersonal chemistry, catalyzed by chemical intoxicants, that comes from in-person socializing in pubs and cafés. The shared experience of music, happy chatter, effortlessly synchronized conversation, rising endorphin levels, and reduced inhibitions is impossible to replace with any technology that we currently possess. The culture that produced Göbekli Tepe would never have gotten off the ground if the awe-inspiring ceremonies conducted there were meant only to be webcast. This has to be taken into account when considering what is lost and what is gained in a world where human interaction more and more takes place only through the medium of our computer screens.

A newspaper editorial[39] published in the early months of 2020, as the Covid-19 pandemic caused governments to begin locking down multiple sectors of public life, observed that the post-coronavirus economy will be a much smaller, less productive one. "In a world where the office is open but the pub is not," it noted, "qualitative differences in the way life feels will be at least as significant as the drop in output." This has certainly been true, but it is equally important to note that the absence of pubs has had a direct effect on both the qualitative feeling of life *and* actual productivity. For instance, I would venture a prediction that, controlling for other factors, innovations or new patents in a given field in the year or two following a canceled in-person annual conference will drop when measured against baseline, and then rise again after in-person socializing and drinking is reestablished. If pubs and cafés are crucibles of innovation, we should see their replacement with sterile teleconferencing causing the same stagnation in collective creativity that followed the imposition of Prohibition in the early-twentieth-century United States.[40]

People working from home, interacting with colleagues only through video chat and email and text, have felt not only more alienated and disconnected, but have also likely experienced fewer creative insights, lacking the stimulus of meandering, unpredictable, and therefore potentially innovative discussions. Video meetings are probably more efficient; but efficiency, the central value of Apollo, is the enemy of disruptive innovation. The pub doesn't just make us feel better; used properly, it makes us, in the long run, work better.

Those aspects of our chimpanzee or wolf nature that keep us narrowly on task and alienated from one another—goal orientation, self-control, self-interest—are most effectively and thoroughly cast aside by throwing people together in person, surrounding them with multi-sensory distractions, and allowing them to gradually but steadily downregulate their PFCs with intoxicants. It would be wonderful if Zoom could create a function that allowed you to synchronize your lighting and music with that of your potential business or research partner's home office, while simultaneously zapping both of you with a powerful transcranial magnet. While we wait for the Zoom development team to implement this upgrade, however, good old-fashioned alcohol, consumed in a relaxed, in-person setting, remains our simplest and most effective cultural technology.

OFFICE PARTIES: PROS AND NOT JUST CONS

Focusing on the ancient social function of intoxication can also help us think more clearly about other aspects of professional life.

Take, for instance, debates about the office party. Faced with the consequences of immoral or even criminal behavior at drunken corporate gatherings, many companies have made, or are considering making, such events completely dry. Beer kegs, complimentary wine, and the cocktail bar are being replaced with sparkling water and spirulina smoothies. Social mixers are also being entirely nixed in favor of structured (and sober) group-building exercises, like escape rooms or laser tag.

The cost-benefit reasoning that drives these choices sees, on one side, the not inconsiderable financial, personal, and potentially legal costs of not only allowing, but actively enabling, employees to get drunk. These are then weighed against—what? Just some fun? When the calculus is framed in this way, fun versus concrete, measurable costs, fun is always going to lose.

Armed with our evolutionary analysis of intoxication, a more accurate way to proceed would be to weigh the undeniable and clear costs of alcohol consumption against its harder-to-perceive benefits. There is a reason that highly successful companies, such as Google, continue to allow alcohol to play a structural role in their institutional life. As we have seen, alcohol enhances creativity, both individual and group. It can help overcome trust issues and allow for a freer exchange of ideas. It is also the case that office parties do, or at least can, build corporate spirit and group bonding. The plot of the rather funny movie *Office Christmas Party* (2016) is motivated by the notion of a high-level investor who is unwilling to take a stake in a tech company because of its lack of cohesive culture, for which said party is proposed as a remedy. (Spoiler alert: It works.) While live reindeer, hard alcohol in the water cooler, and cocaine in the snow machine are probably overkill, these are illustrations of how taking the benefits of intoxication seriously could help tip the scale in favor of fun, and uncover the genuine costs created by taking the party out of the office party.

In an early and influential review of the positive effects of moderate alcohol consumption, the clinical psychologist Cynthia Baum-Baicker concluded that, in line with what we've seen in previous chapters, the primary benefits are a reduction in both physiological and self-reported stress levels, as well as enhanced mood.[41] This conclusion is reflected by higher levels of reported happiness, euphoria, and conviviality, and lower levels of reported tension, depression, and self-consciousness, among subjects who have become inebriated in the lab.[42] In surveys, adults and college students report frequently using alcohol to overcome social anxieties, or as a "shortcut to feelings of similarity, inclusion and belonging,"[43] which accords with classic anthropological work documenting alcohol use around the world to reduce stress in social situations.[44]

Enhanced mood and dampened stress are great for the person who's well into their second complimentary glass of wine, but does this actually lead to group bonding? People become more boisterous and talkative at a .08 percent BAC, but are they actually connecting with others or just enjoying the sound of their own voices? Laboratory work that directly examines the effect of alcohol intoxication on social bonding is sparse,[45] but preliminary data supports the idea that alcohol's effects on mood and anxiety are beneficial for the group as well as the individuals involved. A 2012 study by the psychologist Michael Sayette and colleagues, for instance, involved putting hundreds of subjects, all social drinkers unknown to one another, in social triads where they were asked to converse casually over drinks for half an hour before completing several tasks.[46] While the tasks were the ostensible purpose of the study, the researchers were actually interested in the initial conversations, which were videotaped and analyzed. Subjects in the alcohol condition were given a series of vodka-cranberry cocktails, while a placebo-condition group received virgin cocktails in a glass that had been smeared with a bit of vodka to convince subjects that they were drinking alcohol. A third group was simply given cranberry juice.

Hypothesis- and condition-blind coders were asked to assess individuals' facial expressions as well as group speech patterns during the informal conversations. Bonding was deemed to have occurred in cases where all three participants simultaneously displayed Duchenne, or genuine, smiles (as in Figure 4.2), which can be distinguished from consciously willed smiles. Evenly distributed and sequential turn-taking in conversation was similarly taken as a sign of positive group dynamics.

Figure 4.2. Subjects simultaneously displaying a Duchenne, or genuine, smile.[47]

The subjects themselves were also asked, at the end of the experiment, to rate the degree of bonding in their group by reporting the extent to which they agreed or disagreed with statements such as "I like this group" or "The members of this group are interested in what I have to say." Previous work had established that responses to this Perceived Group Reinforcement Scale (PGRS)[48] correlate with nonverbal measures of social bonding.

The results were clear. "Alcohol consumption," the authors concluded, "enhanced individual and group-level behaviors associated with positive affect, reduced individual-level behaviors associated with negative affect, and elevated self-reported bonding." A later analysis of the social dynamics reflected in the videos found that intoxication enhanced the "contagion" of smiling and positive affect: genuine smiles that popped up in drinking groups were more likely to spread to everyone, rather than simply being ignored. This contagion effect was particularly dramatic in male-dominated groups, where smiles in the placebo and control conditions tended to go unreciprocated.[49] Crucially, these positive effects on group bonding were driven by the pharmacological effects of the alcohol: The placebo group resembled the control group, and both differed significantly from the alcohol group on all measures. One recent summary of research on the effect of alcohol on social cohesion and intimacy concludes that it "can increase self-disclosure, can decrease social anxiety, and consistently increases extraversion, including a gregariousness facet subscale. In addition, research has found that alcohol can increase happiness and sociability, helping behaviors, generosity, and social bonding, and decrease negative emotional responses to social stressors."[50]

This all suggests that replacing alcoholic cocktails with virgin cocktails at the annual office party will defeat much of the purpose of holding such events in the first place. While three college students smiling at one another across a Formica lab table is not as dramatic or entertaining a data point as the actor T. J. Miller launching himself down a slide full of cocaine in a Santa outfit, this study does provide us with actual empirical evidence that alcoholic office parties have positive social functions, in addition to their more obvious costs. Crucially, as this study reveals, it's important to note that the ethanol itself, not just the setting or cultural expectations, is doing the work.

LONG LIVE THE LOCAL

We have noted the function of alcohol in enhancing group creativity or cognitively de-arming potentially hostile people in the contexts of politics, business, and academia. It plays an equally ubiquitous role, of course, in all sorts of low-key, everyday, informal social gatherings. An early Chinese text coined the much-quoted phrase, "The gatherings of all happy events cannot do without wine,"[51] and of course the wine-centered symposium was the model of ancient Greek sociality. The mead hall served as the center of communities in medieval Anglo-Saxon cultures, and continues to do so to this day in the form of the alehouse or the pub.[52] In colonial America, every town had its tavern, which was typically one of the first buildings constructed, right next to the church and/or meetinghouse.[53] From the *salon* of France to the *kabak* of early modern Russia[54] to the saloon of the American frontier, the mild dissociation from reality caused by alcohol and casual, relaxed socializing have been inseparable.[55]

British readers will likely maintain that social drinking has reached the pinnacle of cultural evolution in the form of the pub or "local." A publication by a group of anthropologists and sociologists from 1943, entitled *The Pub and the People,* provides a wonderfully amusing and bewildering account, worthy of Borges, of the variety of activities observed in the 300 or so pubs of Bolton (referred to as "Worktown"), a northern British textile-manufacturing town.

These are the things that people do in pubs:
SIT and/or STAND
DRINK
TALK about betting, sport, work, people, drinking, weather, politics, dirt
SMOKE
SPIT

Many PLAY GAMES
    cards
    dominos
    darts
    quoits

Many BET
    receive and
    pay out losings and winnings

PEOPLE SING AND LISTEN TO SINGING: PLAY THE PIANO AND
LISTEN TO IT BEING PLAYED

THESE THINGS ARE OFTEN CONNECTED WITH PUBS....
    ....weddings and funerals.
    quarrels and fights.
    bowls, fishing and picnics.
    trade unions.
    secret societies. Oddfellows. Buffs.
    religious processions.
    sex.
    getting jobs.
    crime and prostitution.
    dog shows.
    pigeon flying.

PEOPLE SELL AND BUY
    bootlaces, hot pies, black puddings, embrocation....

All of these things don't happen on the same evening, or in the same pubs. But an ordinary evening in an ordinary pub will contain a lot of them.[56]

As Griffith Edwards notes, this study "showed the pub as an institution to be comprehended in terms of multiple functions and symbolisms, rather than just a space where alcohol was sold and consumed." The pub, much like the French café, is a broad tent, welcoming families and old people, bands of drinking friends, lone writers, and couples on dates. The fact that the talking and eating and dart playing and pigeon flying are all gently but insistently lubricated by alcohol is key to the pub's ability to serve as a hub for casual, informal, and spontaneous social interaction. You might get into a quarrel or fight, but you might also join a secret society or pick up a nice set of bootlaces. Just steer clear of the spitting.

A research initiative into the role of the pub in modern British culture is being directed by the anthropologist Robin Dunbar, one of the most active contemporary scholars exploring alcohol's contribution to human sociality. One of the team's findings, from survey data about pub use in Britain, found that people who had a neighborhood pub that they frequented regularly

> had more close friends, felt happier, were more satisfied with their lives, more embedded into their local communities, and more trusting of those around them. Those who never drank did consistently worse on all these criteria, while those who frequented a local did better than regular drinkers who had no local that they visited regularly. A more detailed analysis suggested that it was the frequency of pub visits that lay at the heart of this: it seemed that those who visited the same pub more often were more engaged with, and trusting of, their local community, and as a result they had more friends.[57]

Another survey from the U.K., run on a sample of 2,000 adults[58] and focused on communal eating habits, found that four variables had significant impact on reported feelings of closeness with dinner-mates: number of diners (about four, including the respondent, seemed optimal), presence of laughter, presence of reminiscences, and consumption of alcohol.

As mentioned in Chapter Three, Dunbar attributes this bonding effect to endorphins, which are independently boosted by alcohol and laughing. We can see a kind of virtuous circle here, where alcohol not only itself triggers endorphin release, but by lowering the barrier to laughter and singing and dancing and perhaps more risqué behavior, encourages actions that serve to further ramp up endorphin levels. It is the ideal social enhancer. Reflecting on the uniquely social nature of alcohol, as compared to drugs like cannabis or psychedelics, Dunbar and Kimberley Hockings conclude, "What sets alcohol apart is its use in social contexts rather than for quasi-religious experiences or for purely solitary hedonic pleasure... [It] opens the social pores, allowing more relaxed social interaction, calms the nerves, and creates a sense of community."[59]

This opening of "the social pores" has many socially useful knock-on effects, beyond individual relaxation and group bonding. When venues for alcohol-based socializing disappear or wither away, societies lose not only their centers of communal bonding and good cheer, but also channels for candid exchange and communication. In June of 2018, the impending closure of the Gay Hussar, a venerable and notorious restaurant and drinking hole in London's Soho where politicians, journalists, and union leaders would mix, prompted an article by the journalist Adrian Wooldridge bemoaning the broader demise of the liquid lunch in British politics.[60] It does an excellent job of highlighting what is lost as the influence of alcohol is gradually drained out of public life:

The saddest reason is the rise of a professional political class. Drink provided a link between politics and society. The Labour Party recruited MPs and activists from working men's clubs that existed in large part to provide workers with cheap drink. Ministers routinely let their guard down when they demolished the ministerial drinks cabinet with their civil servants and advisers. Today, both Labour and the Tories recruit their MPs from think-tanks and ministers keep up their guard at all times. The decline of political drinking has snapped yet another link between the political elite and the people that they are supposed to serve.

We will turn later to the noxious role that booze-soaked networking plays in reinforcing old boys' clubs and excluding outsiders. But it is important to note that the demise of informal venues where people can gather, mix, and drink represents a loss of community, honesty, and relationship building that may indeed have negative downstream political and social consequences.

In an important review of the empirical literature on the psychological benefits of moderate alcohol use, addiction researchers Stanton Peele and Archie Brodsky conclude:

> To a greater degree than either abstainers or heavy drinkers, moderate drinkers have been found to experience a sense of psychological, physical, and social well-being; elevated mood; reduced stress (under some circumstances); reduced psychopathology, particularly depression; enhanced sociability and social participation; and higher incomes and less work absence or disability. The elderly often have higher levels of involvement and activity in association with moderate drinking, while often showing better-than-average cognitive functioning following long-term moderate alcohol consumption.[61]

191

In other words, go to the pub and have a pint or two. All things considered—liver damage, calories, and all—a spot of social drinking is good for you, and this has nothing to do with any French paradox or narrow health benefit. Moderate, social drinking brings people together, keeps them connected to their communities, and lubricates the exchange of information and building of networks. We social apes would find it very challenging to do without it, both individually and communally.

BEAUTY IS IN THE EYE OF THE BEER HOLDER: SEX, FRIENDSHIP, AND INTIMACY

The writer Adam Rogers provides a wonderful description of the gentle but insistent arrival of the alcoholic buzz, "the warm, spreading tingle, that slight sense that your brain is still looking at something even after your eyes have moved away. Maybe you're more confident, happier. You were tense; now you're relaxed. Your friends get better looking. Another drink seems like an even better idea."[62] As we have discussed above, alcohol can enhance and ease all manner of social interactions. It is worth turning now to its role in facilitating more personal facets of social life: sex and intimacy.

Among the various theories about alcohol use that have been bandied about in anthropological circles for centuries, perhaps the most idiosyncratic is that of a certain H. H. Hart, who in a 1930 article[63] proposed that alcoholic intoxication serves as a substitute for sexual pleasure because of the decrease in libido it causes. Speak for yourself, Monsieur Hart. The conjunction of alcohol and sex is as ancient as alcohol itself. We noted in the Introduction how the pairing of beer and sex was the catalyst for the taming of the wild man, Enkidu, in ancient Sumerian myth. Cylindrical seals from Mesopotamia dating back as far as the early third millennium BCE frequently depict sexual acts accompanying beer drinking. Piotr Michalowski has observed that, just as the biblical *Song of Songs* declares that love is sweeter than wine, ancient Sumerian poetry celebrates the sexual enjoyment of a goddess as being as "sweet as beer." In an ancient Greek play, *The Bacchae* ("Worshippers of Bacchus"), a herdsman explains the purpose of Dionysian rituals in terms of its facilitation of myriad pleasures, love paramount among them:

> [Dionysus'] powers are manifold;
> But chiefly, as I hear, he gave to men the vine
> To cure their sorrows; and without wine, neither love
> Nor any other pleasure would be left for us.[64]

In Plato's *Symposium,* Socrates prescribes wine for lovemaking. "Wine makes the soul damp, puts trouble to sleep, and wakes kindly feelings," he says, but should be consumed in moderation so that one can "arrive at playfulness."[65]

This link between alcohol and love, wine, and flirty playfulness is still very salient today. There is a massive literature on explicit and implicit attitudes concerning alcohol and sex, the common and cross-cultural view being that alcohol, through its disinhibiting effects, facilitates sexual behavior and enhances sexual experience.[66] Expectancy clearly plays some role: The linkage of alcohol and sex in advertising and media creates a cultural connection that is active in people's minds, distinct from whatever psychological effects ethanol itself might have. The advent of balanced-placebo designs, however, has allowed researchers to disentangle expectancy and pharmacological effects, showing that the latter are also very potent. Indeed, in some studies the most powerful effects were recorded when subjects were falsely led to believe they were not drinking alcohol.[67]

The ancient and widespread idea of alcohol as aphrodisiac is grounded in several of its basic psychoactive effects. Its stimulating function is not limited to general mood: The dopamine boost provided by alcohol directly increases sexual desire in both males and females, from fruit flies to humans.[68] Ironically, its depressant effect simultaneously—and notoriously—impairs actual sexual performance, reducing genital arousal and increasing the time to orgasm in both men and women. (If you are curious about how researchers measure physical arousal, google "penile plethysmograph" or "vaginal photoplethysmography.") Shakespeare's famous dictum that drink "provokes the desire but takes away the performance"[69] thus has some serious empirical evidence behind it.

The notion that intoxication increases the perceived attractiveness of others is also an ancient one. It can be traced all the way back to Aristotle, who noted that "a man in his cups may even be induced to kiss persons whom, because of their appearance or age, nobody at all would kiss when sober."[70] It is also supported by experimental literature. In both laboratory and more naturalistic environments, like pubs or campus parties, heterosexual subjects at a moderate level of inebriation (around 0.08 percent BAC) rate members of the opposite gender as more attractive than sober controls, and the effect is seen in both men and women.[71] Subjects given alcohol rather than a placebo rate photos with sexual content as more appealing and choose to gaze at them longer.[72] Interestingly, the effect is more pronounced in women, which may reflect greater inhibitions created by cultural norms that alcohol then helps to downregulate.[73] As the musician and satirist Kinky Friedman summarizes, "Beauty is in the eye of the beer holder."[74]

Less commonly known is that the alcohol-induced tendency to see others as more attractive, as well as to be less inhibited about sex, is magnified by the fact that mild intoxication also makes one more attractive. When drunk, you physically *appear* more attractive to others: Photos of moderately intoxicated people are rated as more attractive than photos taken of the same person when they are sober.[75] We can see this in Figure 4.3 below. These shots are taken from a wonderful project by the Brazilian photographer Marcos Alberti. As he explains:

> The first picture was taken right away when our guests have just arrived at the studio in order to capture the stress and the fatigue after a full day after working all day long and from also facing rush hour traffic to get here. Only then fun time and my project could begin. At the end of every glass of wine a snapshot, nothing fancy, a face and a wall, 3 times. People from all walks of life, music, art, fashion, dance, architecture, advertising got together for a couple of nights and by the end of the third glass several smiles emerged and many stories were told.

Fig. 4.3. Photographic subjects after one, two, and three glasses of wine (courtesy of Marcos Alberti, The Wine Project: www.masmorrastudio.com/wine-project)[76]

These photos illustrate one aspect of alcohol's attractiveness-enhancing effects: The tense work self is gradually replaced with someone infinitely more relaxed, confident, unselfconscious, and happy.

This effect is then magnified by the fact that you internally *feel* more attractive when drunk, above and beyond whatever external attractiveness boost you might be enjoying, as a result of what is sometimes called the "self-inflation" effect. Drunk subjects rate themselves as more attractive than external observers do, and the more they drink the more attractive they feel.[77] This is due to both mood alteration and cognitive impairment. Dopamine simply makes us feel good—expansive, confident, and friendly. Alcohol-induced cognitive myopia simultaneously reduces self-awareness.[78] This one-two punch makes it more likely that you will, for instance, inflate self-ratings on a personality questionnaire when it comes to qualities where you otherwise feel lacking. If you are worried about not being as intelligent or witty as your peers, a couple drinks will make you a genius comedian, at least in your own eyes.[79] This is all wonderfully satirized in an observation by the ancient Greek philosopher Philostratus, "The river [of wine] makes men rich, and powerful in the assembly, and helpful to their friends, and beautiful and...tall; for when a man has drunk his fill of it he can assemble all these qualities and in his thought make them his own."[80]

As we will discuss in the next chapter, enhanced sexual desire, distorted social perception, and cognitive myopia can easily congeal into a poisonous cocktail, driving seriously bad behavior ranging from poor sexual decisions to sexual harassment and abuse. This is especially the case when the rise in BAC is turbocharged by relatively novel (from an evolutionary perspective) distilled spirits, is socially unregulated, and/or occurs in young people whose partially developed PFCs really don't need any more downregulating. The point here is merely that, when used in moderation by mature, consenting adults, alcohol is a valuable mind-hacking tool and aid to intimacy.

It is hard to imagine a better cultural solution than one or two glasses of wine, combined with a meal, for getting a nervous, budding romantic couple over initial awkwardness or anxiety. Drinks at the end of the day, or beginning of the weekend—what the addiction researchers Christian Müller and Gunter Schumann dryly characterize as a "'scheduled' and time-dependent...transition from professional to private microenvironments"[81]—help uncoupled people meet potential partners and established couples to flip from task-oriented, wolf-mode to relaxed, Labrador-like intimacy. This is why champagne or wine is associated with romantic occasions, like weddings and Valentine's Day. For some reason, one television image that has stuck with me from my youth in South Jersey comes from cheesy advertisements for romantic getaways in the Pocono Mountains of Pennsylvania, which inevitably featured enormous heart-shaped hot tubs, an ice bucket with champagne, and two glasses. Corny, but an accurate depiction of an effective cultural technology.

Turning back to Figure 4.3, it is also worth noting that Alberti's subjects are not arriving at his studio for a blind date, but are rather friends and colleagues. This highlights the fact that boosted mood and cognitive myopia can be tools for enhancing all sorts of intimacy, not only the kind between sexual or potentially sexual couples. My favorite Chinese poet, Tao Yuanming, has a line describing a reunion with a dear friend he hasn't seen in a long time: "Without saying a word, our hearts were drunk; and not from sharing a cup of wine." As Michael Ing observes, "For Tao, friendship is intoxicating, and true friends understand each other without having to say a word. Friendship, like [wine], nullifies the limitations of self and time. It encourages the loss of oneself in another, and heightens an awareness of this other, more communal, self."[82] Although Tao himself attributes his intoxication to friendship, not wine, it is important to realize that this explicit disclaimer is a poetic trick directing our attention to the substance that actually facilitated this meeting of the hearts.

Across the world, survey respondents asked about their motives for drinking put "social enhancement" at the top of their lists.[83] Through its combination of dopamine boost and cognitive myopia, alcohol relaxes inhibitions and social anxiety. This results in not only increased talkativeness, but also a tendency to veer into private or intimate topics.[84] In photo #1, Alberti's photographic subjects are complaining about traffic or reporting on their day at work; by photo #4 they are sharing deep hopes and aspirations, commiserating about failed relationships, or flirting with one another.

Some research suggests that the instrumental use of alcohol might be especially important for introverts or those with social phobias, who strategically use alcohol to effect "a self-induced, time-restricted personality change,"[85] temporarily transforming themselves into extroverts for long enough to make it through a cocktail reception or dinner party.[86] (Introverted readers may well recognize this particular mind hack.) Alcohol also facilitates the perception of positive emotions in faces and enhances empathy, with the effect being stronger for people who tend to be inhibited or low in empathy.[87] Large-scale epidemiological studies show that moderate alcohol consumption, as opposed to both complete abstinence and heavy drinking, is associated with closer friendships and better family support.[88] The social lubrication provided by alcohol is thus crucial not only in helping selfish primates to solve cooperation dilemmas and innovate effectively, but also to forge and maintain intimate personal bonds.

COLLECTIVE EFFERVESCENCE:
TEQUILA SHOTS AND BURNING MAN

Up to this point we've been looking at the social effects of BACs hovering around 0.08 percent, which would be produced by a few glasses of beer or wine consumed leisurely and with some nibbles. What happens, though, when the tequila bottle comes out? The social landscape gets shakier at BACs exceeding .08 or .10 percent, but there is evidence that at least occasional, well-timed excess can in some cases be socially useful.

For instance, there are few organizations as results-driven as the U.S. Navy Seals. It is therefore worth noting that Navy Seal commanders find it useful to, on rare occasions, introduce large quantities of alcohol to their training process in order to build team spirit among their units. As Jamie Wheal and Steven Kotler report in their book *Stealing Fire,* the founder of SEAL Team Six, Richard Marcinko, caps off the grueling experience of boot camp with "a time-tested bonding technique: getting drunk." They note: "Before deployment, he'd take his team out to a local Virginia Beach bar for one final bender. If there were any simmering tensions between members, they'd invariably come out after a few drinks. By morning, the men might be nursing headaches, but they'd be straight with each other and ready to function as a seamless unit."[89] This belief in the power of occasional "benders" jibes with survey data from college students showing that those who experienced occasional blowout drinking sessions had deeper social ties, measured in terms of greater intimacy and disclosure with friends and romantic partners, than both those who always drank moderately and those who binged frequently. As the authors noted, "It appears that Disraeli's observation that 'There is moderation even in excess' provides an appropriate context for the present results."[90]

199

Perhaps the most epic bender one can imagine is the annual, weeklong event called Burning Man, held in the Black Rock Desert of Western Nevada. Burning Man is probably the closest experience a modern person will have to ancient Dionysian revels, an exhilarating, buzzing confusion of heat and dust, art and sex, music and dancing, mutant vehicles and wild outfits, social activism and experiments in collective living, all fueled by startling quantities of alcohol, psychedelics, and uppers, and heightened by severe sleep deprivation. As the sociologist Fred Turner notes, going to Burning Man has become something of a rite of passage among tech and information industry workers in Silicon Valley. In 1999, Google's founders, Larry Page and Sergey Brin, featured the Burning Man logo on the Google home page to signal to the world that they (along with many of their staff) were headed to the festival, and CEO Eric Schmidt was reportedly hired because he was also a "Burner." Collective experience of the intensity and physical discomfort of Burning Man is seen in the Valley as a way of building internal cohesion and culture. As Turner notes:

> Larry Harvey [one of the founders of the original Burning Man event in San Francisco] explained that the world outside Black Rock City was "based on separating people in order to market to them." At Burning Man, he argued, participants would encounter the "immediacy" of art and through it, ecstatic feelings of community. In that sense, he implied that Burning Man would offer its participants the feeling of "effervescence" that Durkheim long ago argued formed the basis of religious feeling. Gathered in the desert, participants in the festival can feel an electric sense of personal and collective transformation.[91]

Some companies have even taken a shot at creating smaller versions of Burning Man in their own work retreats, aiming to capture that same sense of "group flow." "One CEO of a small [tech] start-up," Emma Hogan reports, "describes how, on an away-day with his company, everyone took magic mushrooms. It allowed them to 'drop the barriers that would typically exist in an office,' have 'heart to hearts,' and helped build the 'culture' of the company."[92]

Of course, the need for ecstatic experience and group bonding goes far beyond the business world. Larry Harvey's comments about modern society being intent on "separating people in order to market to them" recalls the author Barbara Ehrenreich's biting observation that missionaries sent to South Africa deliberately focused on repressing traditional dances as a means for "weakening the communistic relations of members of a tribe among one another," with the goal of, as she sarcastically puts it, "letting in the fresh, stimulating breath of healthy individualistic competition."[93] As we have noted, many scholars of ritual and religion have highlighted the role of "techniques of ecstasy" (Mircea Eliade) for bonding individuals together, or characterized group dance as the "biotechnology of group formation" (Robin Dunbar). As Ehrenreich notes, group dancing is one of the earliest human-centered scenes found in pre-historical art from around the world, which is puzzling considering the fact that ecstatic dancing seems like "a gratuitous waste of energy" when compared to such practical and essential activities as hunting, gathering, cooking, or making clothes.[94] To have remained a universal and basic human practice for so long, it must be paying for itself, as it were, in evolutionary terms.

Part of the payoff relates to individual mental health, the need we have for a "vacation from the self," as we'll discuss in the next section. Clearly, though, an important function of "dancing in the streets" is the creation of group identity, which is as essential as water and food to our hyper-social and vulnerable species. As the psychologist Jonathan Haidt and colleagues have noted, group synchrony and rhythm help to foster the feeling of being part of a "hive mind" that we odd primates have come to crave. A desire for ecstatic experiences evolved and survived because it was adaptive for humans—those who danced together worked and fought together, and learned to trust and be trusted.[95] It is true that, in the industrialized world, modern institutions and rule of law have mostly replaced religious ritual and ecstatic bonding as cultural technologies for ensuring cooperation. This relatively recent development does not, however, immediately eliminate such a deep and basic drive from our motivational systems. People vote and pay taxes, but they still want to dance.

Do we need to be drunk in order to dance? Classic work on the anthropology of ritual and religion tends to give short shrift to chemical intoxication. For instance, in his famous description of collective effervescence, Émile Durkheim begins his list of behaviors that lead to ecstatic experience with "cries, songs, music, violent movements, and dances." Only at the end does he add, "the search for exciteants which raise the vital level, etc."[96] This is an oddly coy and vague reference to, presumably, alcohol and other drugs. In his landmark (and enormous) work, *Ritual and Religion in the Making of Humanity*, Roy Rappaport only mentions "drug ingestion" a few times, in passing, and typically at the end of a long list of other, presumably more important, features of ritual such as "length...tempo, unison, density of symbolic, iconic and indexical representation, sensory loading, strangeness...or pain."[97] This curious reluctance of anthropologists and historians of religion to focus on chemical intoxicant use makes it difficult to assess how crucial it is to group ecstasy. This is where investigating modern gatherings can be useful for teasing apart the relative contributions of various techniques, especially because it is difficult to obtain permission to get human subjects in the lab drunk or seriously lit up on LSD.

If we look at Durkheim's or Rappaport's list of techniques for achieving ecstatic bonding, they seem to describe pretty well not only Burning Man, but any of a number of group events that take place periodically around the world, including music festivals, modern shamanistic gatherings, and raves. Although chemical intoxicant use is an assumed ingredient in these events, the degree to which the drugs themselves are responsible for producing Durkheimian effervescence and group bonding remains an open question. A recent study of multiday mass gathering events in the U.S. and the U.K.—specifically outdoor festivals and concerts that mingled music, dancing, synchrony, and lots of drugs—took an initial step in the direction of untangling these factors. The researchers went on-site and interviewed over 1,200 attendees about the nature and quality of their experiences, as well as their recent psychoactive drug use. They found that drug use—particularly the use of psychedelics and benzos, such as Valium—correlated with an increased likelihood to report that the event was accompanied by positive mood, involved social connection, and was a transformative experience.[98] Dancing and listening to music was fine, but drugs appeared to provide the catalyst for transformation and bonding. This study thus provides some very preliminary evidence that chemical intoxicants play a crucial, and too often unmentioned, role in satisfying a basic human need for ecstasy.

In the international surveys about alcohol use mentioned above, social enhancement tends to be the most frequently cited motivation for drinking. In a close second place, however, is something researchers rather blandly refer to as "self enhancement (internal and positive emotions)"—in other words, having fun.[99] In light of that, let's conclude our discussion of intoxication's contribution to contemporary life by addressing two fundamental goods that tend to be marginalized in Puritanical academic and public discourse: ecstasy and pleasure.

ECSTASY: VACATION FROM THE SELF

Man, being reasonable, must get drunk.

—Lord Byron[100]

Much of this book has been dedicated to the individual and social functions of moderate intoxication. As the writer Stuart Walton notes, however, "Moderation is not, in fact, an ideal that finds much house-room within the domain of intoxication. Indeed, intoxication is in itself the opportunity for a temporary escape from the moderation that the rest of life is necessarily mortgaged to." This echoes the sentiments of that great defender of Dionysus against Apollonian control, Friedrich Nietzsche,[101] who noted that Dionysian *orgeia*—all-night, alcohol- and dance-fueled revels that gave us our word "orgy"—were intended to provide a release from "the horror of individual existence" into the "blissful ecstasy" of "mystical Oneness."

> Under the charm of the Dionysian not only is the union between man and man reaffirmed, but nature which has become alienated, hostile, or subjugated, celebrates once more her reconciliation with her lost son, man...Now the slave is a free man; now all the rigid hostile barriers that necessity, caprice, or impudent convention have fixed between man and man are broken. Now, with the gospel of universal harmony, each one feels himself not only united, reconciled, and fused with his neighbour, but as one with him, as if the veil of *māyā* had been torn aside and were now merely fluttering in tatters before the mysterious primordial unity.[102]

For the life-affirming acolyte of Dionysus, sensual pleasure is "free to free hearts, the earth's garden-joy, an overflowing of thanks," the "great restorative and reverently-preserved wine of wines."[103]

Another eloquent advocate of drunken, ecstatic unity with the universe was the Chinese poet Liu Ling (c. 221 to 300 CE). Liu was a notorious drinker who would spit on any talk of moderation. During one of his benders he reportedly got stark naked in a public-facing room of his house, drawing rebukes from passersby. He is said to have shouted at them, "For me, Heaven and Earth are the rafters and roof of my house; this room is just the trousers of my garment. What are you gentlemen doing inside my trousers?"[104] He expands upon this theme in his famous, "Celebration of the Power of Wine":

There was a Great Man:
For him all Heaven and Earth were in a day,
A myriad ages passing in an instant.
The Sun and Moon were his door and window. . . .
His happiness was overflowing.
At times he was drunk,
And at times he would sober up.
He listens so quietly he could not hear the sound of thunder,
Looks so closely he could not see the mass of Mount Tai.
He would not notice the chill or heat cutting his skin,
Nor feel any emotion of interest or desire.
He looks down at the profusion of the myriad things,
Like duckweed floating in the Jiang and Han rivers.[105]

As Michael Ing observes, for early Chinese poets and writers, wine taken in this kind of excess "is a sacred drink—immortalizing and sanctifying those who consume it" by eliminating the fetters of the individual self, allowing a communing with the universe as a whole.[106]

We don't have to share these refined metaphysical ambitions to understand the draw of ecstatic intoxication. The need for ecstasy, like the desire for play, is basic to humans, and seems to be shared by other species as well. Lots of animals play, and a significant number of animals have been observed getting completely out of their minds on drugs. From stoner dolphins copping a high off puffer fish toxin to lemurs blowing their minds with toxic centipedes,[107] the ubiquity of chemical intoxicant use throughout the animal world has led the psychologist Ronald Siegel to declare that "intoxication is the fourth drive," after food, sex, and sleep.[108]

Humans, however, are even more in need of ecstasy than most. We are afflicted by a malady that, as far as we know, is not shared with any other species: conscious self-awareness. As Albert Camus once observed, in his reflections on the Sisyphean nature of human existence, "If I were a cat among animals, this life would have meaning, or rather this problem would not arise, for I should belong to this world. I would *be* this world to which I am now opposed by my whole consciousness."[109] One of the primary functions of alcohol and other chemical intoxicants is to, at least temporarily, abolish what the social psychologist Mark Leary has called the "curse of the self," our goal-oriented, anxiety-prone inner color commentator who is always getting in the way of our ability to simply *be* and enjoy the world. "Had the human self been installed with a mute button or off switch," Leary writes, "the self would not be the curse to happiness that it often is."[110] Human selves do not, in fact, come pre-installed with a mute button, which is precisely why we reach for the bottle or joint. "We now spend a good deal more on drink and smoke than we spend on education," Aldous Huxley observes, because "the urge to escape from selfhood and the environment is in almost everyone all the time."[111] This urge finds its outlet in spiritual practices, like prayer, meditation, or yoga, and also in our drive to drink and get high.

In *Dancing in the Streets,* Barbara Ehrenreich argues that collective, ecstatic rituals have traditionally served to inject regular doses of this crucial Dionysian element into people's everyday lives. From the ancient Greek Dionysian festivals themselves to medieval European carnivals to early American revivalist gatherings, they have provided a liminal space in which people can achieve a degree of ecstasy incompatible with daily routine, something far beyond the tame sociality of the pub or symposium. As we have seen, such spaces still exist in the form of festivals such as Burning Man, and in the scattered continuation of carnivals in the modern world, like Mardi Gras in New Orleans. Less formally, they also survive as raves in all of their various forms, such as the modern Australian "doof," which is held in an outdoor, remote place and involves prolonged vigorous dancing, trance-inducing music, and healthy doses of alcohol, LSD, and MDMA.

Psychedelics, often the drug of choice at these gatherings, are capable of producing incredibly powerful experiences of bliss and rapture. It is too often the case that listening to accounts of other people's trips is as tedious and unhelpful as hearing about what they dreamt last night (or reading a twenty-page treatise on how Truth Is the Color Blue). But there *are* some first-person accounts that manage to convey, as much as the pale medium of words can manage, something of the magic of a psychedelic experience. Alexander Shulgin, a pioneer in the research of synthetic psychoactive drugs, gives this account of his experience on 120 mg of pure MDMA:

> I felt that I wanted to go back, but I knew there was no turning back. Then the fear started to leave me, and I could try taking little baby steps, like taking first steps after being reborn. The woodpile is so beautiful, about all the joy and beauty that I can stand. I am afraid to turn around and face the mountains,

for fear they will overpower me. But I did look, and I am astounded. Everyone must get to experience a profound state like this. I feel totally peaceful. I have lived all my life to get here, and I feel I have come home. I am complete."[112]

Unlike my ill-fated essay on the color blue, which notably failed to revolutionize the intellectual landscape of modern Western philosophy, insights derived from chemically assisted, ecstatic experiences can have lasting effects on everyday life. For instance, there is good empirical data on psychedelic experiences and long-term positive mental health outcomes. Beginning with the famous Good Friday Experiment, in which a bunch of straight-laced divinity school students had their minds blown by 30 mg of purified psilocybin and then followed over the course of twenty-five years, a growing body of evidence suggests that even a single intense experience of chemical ecstasy can provide long-lasting benefits, alleviating depression and enhancing openness to experience, mood, aesthetic appreciation of the world, compassion, and altruistic behavior.[113] One study reported that 67 percent of the subjects who had taken a dose of purified psilocybin considered it either the single most significant experience of their life or in the top five, with many comparing it to the birth of their first child or death of a parent.[114]

Inspired by this sort of research, as well as venerable shamanistic practices, traditional psychedelics such as ayahuasca, psilocybin, and mescaline are now being used to treat addiction, obsessive-compulsive disorder, severe depression, and end of life anxiety.[115] Some of this renewed interest in traditional healing techniques can seem annoyingly trendy or facile: There is no shortage of individual spiritual seekers irritating the locals by showing up in the Amazonian rain forest demanding to score some ayahuasca from a shaman, and there are now even companies offering "psychedelic tourism."[116] Large-scale, and especially longitudinal, data on the effectiveness of these sorts of treatments is still being gathered, but at least one study of ayahuasca use in treating addiction found important positive effects. It did note, however, that the full efficacy of the treatment comes from embedding the drug use within a traditional ritual and symbolic framework.[117] This makes sense in light of empirical literature, as well as experienced users' folk wisdom, which suggests that the conceptual framing ("set") and immediate environment ("setting") are very important in shaping the content and emotional valence of psychedelic trips.

In the 1960s, the anthropologist Douglas Sharon explored mescaline-assisted practices in Peru under the guidance of a local *curandero,* or healer, who explained to him the power of psychedelics in terms that should sound familiar to us by now:

The subconscious is a superior part (of humans)...a kind of bag where the individual has stored all his memories, all his valuations...One must try...to make the individual "jump out" of his conscious mind. That is the principal task of *curanderismo.* By means of the magical plants and the chants and the search for the roots of the problem, the subconscious of the individual is opened like a flower, and it releases these blockages. All by itself it tells things. A very practical manner...which was known to the ancients [of Peru].[118]

Getting the afflicted person to "jump out" of their conscious mind is one way to describe the effect of dialing down the PFC. Like the rats in the experiments with overcrowded cages, humans in civilization live squeezed together, constantly rubbing shoulders with strangers, in a way that goes fundamentally against our chimpanzee nature. We delay gratification, accept complex suboptimal compromises, work long days at boring jobs, and endure tedious meetings. We are particularly in need of having our unconscious "opened like a flower"—at least on occasion.

Something is lost in a world where at least a temporary dissolution of the self is never given space to occur. Traditional, psychedelic-assisted healing practices have been one way to facilitate this. As Ehrenreich argues, periodic festivals and carnivals have been another. One of the worries she expresses in her book is that, under the pernicious influence of Apollo, these opportunities for both individual and group ecstatic joy are being squeezed out of our lives in the name of efficiency, health, or morality. The grim wolf has done an effective job of herding the Labrador into line. This view is echoed by those who've observed that religious life over the past couple centuries has seen the gradual replacement of collective bonding with passive, isolated individualism. As Stuart Walton observes about many forms of post-Temperance Christianity:

> Sitting in of an evening, meditating on the Bible, whittling away at a piece of wood or spinning yarn for knitting, was preferable to taking alcohol in groups. In this way, the campaign against intoxication succeeded in atomizing individuals, a move that many of the mass leisure pursuits of the twentieth century would reinforce by encouraging them to combine only in order to stare in ordered passivity at some entertainment spectacle, whether in the cinema, concert-hall, football ground or in virtual reality, whereas intoxication had brought them together in interacting, dynamic gatherings.[119]

Where we would once assemble periodically in mass ecstatic festivals, or at least on a regular basis in pubs for some loose conversation and informal play, the disciplines of modern life have too often channeled our limited leisure time into isolated, homebound activities, like TV watching and video games. The advent of the internet has only made this worse, with social media addiction and relentless torrents of email and texts keeping us on our couches or prone in bed, glued to our individual screens.

The appeal of the wild and sacred, the rush of ecstasy and power, is to my mind best captured in Coleridge's "Kubla Khan":

> And all should cry, Beware! Beware!
> His flashing eyes, his floating hair!
> Weave a circle round him thrice,
> And close your eyes with holy dread,
> For he on honey-dew hath fed,
> And drunk the milk of Paradise.[120]

Ecstasy, incarnate, seems rather far removed from suburban cul-de-sacs and living rooms bathed in a steady glow of TV light or the flickering of tiny smartphone screens. Not much room there for holy dread or imbibing the milk of Paradise. This is a shame, and not just because the milk of Paradise might enhance our HDL levels or purify our water. We are the creative ape, as well as cultural and communal. A fourth C might be thrown in for good measure: humans are also the *conscious* ape. A being that is self-aware, cut off from the undifferentiated, animal flow of experience by the curse of the self, requires release. This brings us to this chapter's final topic: a defense of pleasure for pleasure's sake.

IT'S ONLY ROCK-N-ROLL: DEFENDING THE HEDONISTIC BODY

Demonized from the early modern era well into the twentieth century as the "poisonous tap-root" of all evil,[121] alcohol won back some utilitarian respectability with research suggesting that moderate alcohol consumption—on the order of one to two drinks a day—might reduce risks of heart disease, diabetes, or strokes. As we have noted, though, practicing physicians have never been terribly impressed by this body of research, and have resisted actively recommending light drinking in the same way they do, for instance, regular exercise. The health-based defense of alcohol finally suffered a massive body blow from the 2018 *Lancet* article that has haunted our discussion, a terrible document that concluded definitively that the only safe level of alcohol consumption was zero.

As mentioned above, responses to the *Lancet* study ranged from a predictable "I told you so" from the teetotaler crowd to those wanting to challenge the methodology and salvage some health benefits for alcohol. An alternative tack is the one taken in this chapter: uncovering or drawing attention to the various ways in which alcohol continues to serve important individual and social functions, the value of which must be weighed against the more obvious health risks. Other defenders of continued alcohol use, however, have no interest in discussing costs and benefits, of weighing the one against the other. Their response might be dubbed the It's Only Rock-n-Roll (But I Like It) position. Maybe alcohol is bad for you, maybe it harms society, but I *like* it, it makes me feel good. Plus, proponents typically add, lots of stuff is not good for you, but we do those things anyway because they are *fun*.

As a philosophical hedonist, I am fully on board with the rock-n-roll defenders. It is true that, in our current age of neo-prohibition and general queasiness about risk, we desperately need to be clear about the simple joy of feeling good. In defending the functions of intoxicant use, let us never lose sight of one of the greatest contributions of intoxicants to human life: sheer hedonic pleasure. As Stuart Walton observes in his brilliant, wickedly funny cultural history of intoxication, *Out of It,*[122] "There is a sedimentary layer of apologetics, of bashful, tittering euphemism, at the bottom of all talk about alcohol as an intoxicant that was laid down in the nineteenth century, which not even the liberal revolution of the 1960s quite managed to dislodge." It is worth quoting at length his diatribe against the whiff of Victorian hypocrisy that seems to invariably accompany any discussion of alcohol:

> A hysterical editorial in a tabloid newspaper calling for drinks companies to be made to pay the medical expenses of cirrhosis patients may simply be called the mood-music of the new repression, but how to react to this introductory comment in a monumental history of winemaking by one of its most elegant chroniclers, Hugh Johnson? "It was not the subtle bouquet of wine, or a lingering aftertaste of violets and raspberries, that first caught the attention of our ancestors. It was, I'm afraid, its effect." Quite so, but why the deprecatory mumble? What is there to be "afraid" of in acknowledging that wine's parentage lies in alcohol, that our ancestors were attracted to it because the first experience of inebriation was like nothing else in the phenomenal world? And what else in it attracts the oenophile of tomorrow in the first place, if not the fact that she found it a pleasant way of getting intoxicated today? Can we not say these things out loud, as if we were adults whose lives were already chock-full of sensory experience?

We cannot, he concludes. "It is in many ways easier to be frank today about one's sexual habits than it is to talk about what intoxicants one uses...rendering us all shame-faced inarticulates on the subject."[123] It is time for us to become articulate again. While it is socially acceptable to talk in purely aesthetic terms about our interest in fine wine, microbrewed beer, or designer cannabis, we remain uncomfortable talking about our need for embodied pleasure for its own sake, rather than as a side effect of more respectable, abstract connoisseurship. This is a hang-up that we need to get over.

People masturbate, and they like to get high. We need to be as clear-sighted and free of squeamishness talking about the latter as the former. As the anthropologist Dwight Heath complains, "A fundamental curiosity about most of what is written about beverage alcohol, especially by scientists, health professionals, and other researchers, is that so little acknowledgement is made that the great majority of people who drink do so because they find it enjoyable and pleasurable."[124] In his literary history of drinking, Marty Roth attributes this odd lacuna—one we have also observed in the anthropology and cognitive science of ritual—to a shift in our view of alcohol that can be dated to the mid-nineteenth century. Once simply an assumed component of the good life, a substance "that elevated and released," alcohol came to be seen exclusively through a medicalized lens of addiction and public health impacts. He quotes from an essay by the Spanish philosopher José Ortega y Gasset on the bacchanal paintings of Titian and Velázquez:

Once, long before wine became an administrative problem, Bacchus was a god, wine was divine…Yet our solution is symptomatic of the dullness of our age, its administrative hypertrophy, its morbidly cautious preoccupation with today's trivia and tomorrow's problems, its total lack of the heroic spirit. Who has now a gaze penetrating enough to see beyond alcoholism—a mountain of printed papers loaded with statistics—to the simple image of twining vine-tendrils and broad clusters of grapes pierced by the golden arrows of the sun.[125]

It is of course ironic that this book, while defending the power of Dionysus, has mostly done so in a way that bows to Apollo, the dull god of "administrative hypertrophy." We've spent the bulk of our discussion focused on the practical benefits and uses of alcohol and other intoxicants. It is important not to forget the deeper meaning of simple but powerful images, such as the scene from Titian's *Bacchanal of the Andrians* (1523 to 1526) that graces the cover of this book. As Stuart Walton complains about the back-and-forth debate on topics like the health benefits of wine drinking:

> What all of this ignores is that intoxication is its own justification. We overlook the deleterious effect that drinking may cumulatively have on our livers to the same degree that we might think it nicely reassuring, but no more, to hear that our low-density cholesterol is being kicked into touch by it. Whatever other physiological processes are going on while we drink, our brains are experiencing intoxication symptoms, and the pleasure, satisfaction and relief that that affords were the reasons we scrabbled through the drawer for the corkscrew in the first place.[126]

Morbid cautiousness, and an allergy to pleasure, is, however, not an affliction exclusive to our modern age. Apollo has been with us from the beginning. The danger of losing a feel for Dionysus through excessive Apollonian functionalism is well illustrated in an analysis of a passage from a Greek text dating to the early centuries CE[127] by the philosopher Jan Szaif. It describes the behavior of the virtuous person, who prefers to "engage both in noble actions and in the study of fine/noble things." Because of his social nature, the virtuous person "will also marry, beget children, engage in civic matters, fall in love in the manner of temperate love (*erōs*), and get himself drunk in the context of social gatherings—although not in a primary way." This is an odd locution. What could it possibly mean to get drunk, but not "in a primary way"? Szaif explains:

> As employed in this passage, the term indicates that the activity in question is not chosen because the virtuous person finds it desirable as such, but because of the circumstances or because of certain hypothetical necessities. In this specific case, virtuous people engage in drinking not because drunkenness attracts them, but because it is a concomitant of something else they care for, namely, certain desirable social activities that involve heavy drinking. Yet virtuous people will not engage in these activities for the sake of intoxication, but in order to partake in the communal life, as part of the fulfillment of their social nature.[128]

We can applaud this attitude. Most of this book has focused on precisely this project of unpacking alcohol's functional role in human cultural life. We are an odd, sad species of ape, trying to make our way in societies organized on a scale we are not genetically equipped to negotiate. The discovery of a liquid neurotoxin that helps us to be more creative, culturally connected, and communally trusting was a crucial stage in our evolution, and we need to better understand how intoxicants continue to function for us today. But let us never lose sight of the fact that drinking, or smoking, or taking an occasional mushroom trip is primordially, atavistically *fun*. Let us flash our eyes and drink the milk of Paradise. Let us not be afraid to get drunk "in a primary way," for this is what reconnects us to the flow of experience that other animals get to simply take for granted.

IT IS TIME TO BE DRUNK

Sir David Spiegelhalter, Winton Professor for the Public Under-standing of Risk at the University of Cambridge, disputed the conclusions of the *Lancet* article's authors, noting that the data showed only a very low level of harm in moderate drinkers. "Given the pleasure presumably associated with moderate drink-ing, claiming there is no 'safe' level does not seem an argument for abstention," he said. "There is no safe level of driving, but governments do not recommend that people avoid driving. Come to think of it, there is no safe level of living, but nobody would recommend abstention."[129]

Dry British wit at its best. We should point out, though, that the government does not recommend against people driving be-cause there are clear, observable benefits to driving, which we can weigh against the equally obvious costs. Living is unavoidable—the Rolling Stones inexplicably carry on. Alcohol, on the other hand, stands defenseless against bureaucrats, physicians, and government policy makers in large part because we have failed to do the work of uncovering the evolutionary rationales for its role as the king of intoxicants and unpacking its continued benefits to individuals and societies. Pleasure alone is, alas, rarely broad enough cover for intoxication.

The psychologists Christian Müller and Gunter Schumann make two important points in their review article on "drug instrumen-talization," or the rational, strategic use of chemical intoxicants to achieve specific, desirable outcomes.[130] The first is that, despite justified concerns about alcoholism and drug addiction, the vast majority of psychoactive drug users, across the world and through all age groups, are not addicts, and are at very low risk of becoming addicts.[131] Most people use intoxicants simply as tools for creating desired short-term psychological shifts, in the same way they use coffee to wake up and focus, or a silly TV show in the evening to decompress from a hard day at work. The authors further argue that modern, industrialized societies involve a much higher density and variety of social "microenvironments," to which people are constantly required to adapt, than traditional agricultural, or pre-agricultural, societies. We work from home, collaborate online, network over meals and at receptions, wedge exercise or a stolen moment with our children in between a conference call and a team brainstorming session. Psychoactives—not only pure stimulants like coffee and nicotine, but intoxicants like alcohol and cannabis—might therefore be even more important for us today than they have been historically. Chemical intoxicants may have lured early hunter-gatherers into the agricultural life, and then served as a crucial tool for allowing them to adapt to it. Despite the other tools we currently have at our disposal, we descendants of those first domesticated apes may require chemical support now more than ever.

I would submit that one reason we have trouble properly valuing the benefits derived from chemical intoxicants is because of a false, but deeply seated, dualism between mind and body that colors our judgment. We have no problem with people altering their mood by watching fluff TV or going for a jog, but grow uncomfortable when their psychoactive hack involves a corkscrew and chilled bottle of Chardonnay. A person who meditates for an hour and achieves x percent reduction in stress and experiences a y percent rise in mood is viewed in a much more positive light than one who spent that hour achieving precisely the same results by downing a couple pints of beer. Some of the variance here can be explained by the potential negatives that accompany alcohol consumption— potential for addiction, truckload of calories, damage to the liver— but this is only part of the story.

The prejudice against chemical intoxication is deeply seated not only in our popular consciousness, but also in the scholarly treatment of religion and ritual. We have already noted how the vast majority of literature on ritual bonding and collective effervescence focuses solely on the wholesome dancing and singing, remaining oddly—one might say, Puritanically—silent on the equally pervasive sipping, smoking, and ingesting. Mircea Eliade, in his landmark comparative study of shamanism, famously dismissed drug-induced shamanistic experiences as a "mechanical and corrupt method of reproducing 'ecstasy,'" a "vulgar substitute for 'pure' trance."[132] Similarly, references to chemical intoxication in poetry and literature are often glossed over as simple metaphors. As Marty Roth trenchantly observes, "In commentary on Persian poetry, intoxication is vaporized into allegory, transformed into religious ecstasy before it can even wet its lips."[133]

Aldous Huxley's defense of chemically induced spiritual experience is, in this context, apropos, and may help us to think more clearly about the topic:

> Those who are offended by the idea that the swallowing of a pill may contribute to a genuinely religious experience should remember that all the standard mortifications—fasting, voluntary sleeplessness and self-torture—inflicted upon themselves by the ascetics of every religion for the purpose of acquiring merit, are also, like the mind-changing drugs, powerful devices for altering the chemistry of the body in general and nervous system in particular....[134]
>
> God, [one might insist], is a spirit and is to be worshiped in spirit. Therefore an experience which is chemically conditioned cannot be an experience of the divine. But, in some way or another, *all* of our experiences are chemically conditioned, and if we imagine that some of them are purely "spiritual,"

purely "intellectual," purely "aesthetic," it is merely because we have never troubled to investigate the internal chemical environment at the moment of their occurrence.[135]

Indeed. Recognizing that *all* of our experiences are chemically conditioned might help us to become a little less smug about spending our afternoons meditating or praying rather than sipping some wine in the company of friends or surrounded by the pleasure of a garden. It is our deeply seated, and so typically invisible, mind-body dualism that causes us to systematically, and unfairly, denigrate the role of chemical intoxication in any vision of the good life.

Tao Yuanming produced an immense body of evocative verse on the beauty of nature, the joys of country life, and—not incidentally—the power of wine. His series of poems on the theme of "Returning Home" includes the lines:

Holding the children by the hand, I enter the house,
Where there is wine filling a jug.
Taking the jug and a cup, I pour for myself;
A sideways glance at the courtyard trees warms my face.[136]

As the literary scholar Charles Kwong notes, wine is portrayed here, and elsewhere in the work of Tao Yuanming, as something that is woven inextricably into a vision of the good life:

Enjoyed at home amid the compensating delight of family and Nature, at times in the company of fellow farmers, wine is savoured together with sweaty labour, well-earned leisure, domestic joy and neighbourly affection. Sparkling in the primitive light at dusk and an enlightening [sense of natural-ness], wine has gone far beyond its material properties; it is blended into a rewarding life of self-discovery and self-reliance, into a spirit of rustic simplicity living in existential peace.[137]

It makes intuitive sense, both in early China and the modern world, to talk about wine going "far beyond its material properties" in order to become part of a rewarding, spiritual lifestyle. It is helpful, though, to put aside our intuitive dualism for a moment and see how the warm flush suffusing the poet's nervous system as the first molecules of ethanol begin their work on neurotransmitters in his brain, his cultural expectations about the solace provided by a sip of long-desired local wine, and the pleasure derived from being reunited with his family and neighbors and garden are all simply different facets of a single, physical reality. Let us, like Tao Yuanming, learn to celebrate them all equally.

Given the title of this chapter, it would be fitting to close by citing Charles Baudelaire's famous ode to intoxication, "Get Drunk":

You must always be drunk. Everything depends upon getting drunk, it is all that matters. In order to not feel the horrible burden of time that breaks your back and pushes you to the ground, you must be drunk, perpetually drunk.

Drunk on what? On wine, on poetry or virtue, whatever your taste. But get drunk.

And if sometimes, on the steps of a palace, the green grass of a ditch, or in the dreary solitude of your room, you find yourself sobering up, your drunkenness diminished or entirely gone, and you ask of the wind, of the wave, of the star, of the bird, of the clock—of everything that exists, everything that moans, everything that rolls, everything that sings, everything that speaks—what time it is, the wind, the wave, the star, the bird, the clock will all answer you: "It is time to get drunk! So as to no longer be the martyred slave of Time, get drunk, be constantly drunk! On wine, on poetry, or on virtue, whatever your taste.[138]

We need to be drunk, but upon what? Which facet of material reality should we allow to transport us away from ourselves? It is hard to deny that we humans, to function and flourish, require some degree of intoxication in our lives.

It is quite possible, however, that there are times when meditation, poetry, or virtue might be better options than wine, beer, or Scotch. Any defense of chemical intoxication, the physical rush of ethanol hitting the brain-body barrier, needs to also acknowledge the chaotic and dangerous aspects of bacchanalian excess. Let us therefore turn now to consider a set of serious concerns—some old, some new—about allowing our PFC-downregulation to be performed by chemicals that we drink, ingest, or smoke. Dionysus may have led us, dancing to his pipes, into civilization. But he can also, if we are not careful, turn us into animals.

CHAPTER FIVE

THE DARK SIDE OF DIONYSUS

Wine was central to early Chinese diplomacy and religion and was the basic foundation for community, both among people in this world and between this world and the next. Simultaneously, however, alcohol was also seen as one of the paradigmatic threats to this good order.[1] The evil last rulers of any fallen regime were typically portrayed as drunkards and womanizers who neglected affairs of state and the welfare of the people to debauch with their concubines in artificial lakes full of booze. These stories have their modern counterparts in corrupt Chinese Communist Party officials, who are typically found to be maintaining several mistresses and drowning themselves in rivers of first-growth Bordeaux.

We see similarly ambiguous attitudes toward alcohol wherever we find people drinking. Dionysus was the ancient Greek god of wine, but also of chaos and disorder. His followers achieved exalted states of union with the god, but if you ran across them in the woods they might fall upon you and tear you limb from limb. Throughout the world's ancient literatures, from early China to ancient Egypt and Mesopotamia, to the Hebrew Bible and the New Testament, we find warnings about the dangers of drinking, particularly when indulged to excess. After emerging from the ark, post-flood, with his family and his cargo of animals, the first thing Noah did was build an altar and sacrifice to God in thanks. Religious obligations out of the way, the very *next* thing he did was plant a vineyard and then partake of its fruits (presumably he planted very quick-maturing varietals.) The third thing he did was get so thoroughly wasted that he fell asleep naked, which led to some serious embarrassment on the part of his sons, an awkward attempt to remedy the situation, and, when he finally awoke, an angry curse from Noah, which resulted in an entire line of his descendants being doomed for eternity.[2]

Acquiring wine is an urgent priority, but consuming it can precipitate disaster. An Aztec emperor, in a proclamation issued upon his accession to the throne, similarly warned against the dangers of *pulque:* "What I principally command is that you shun drunkenness, that you do not drink *pulque,* because it is like henbane which removes man from his reason…This *pulque* and drunkenness is the cause of all discord and dissension, and of all revolts and unrest among the towns and kingdoms; it is like a whirlwind that upsets and disturbs everything; it is like an infernal storm that brings with it all possible evils."[3]

Dwight Heath notes that in cultures across the world the use of alcohol is inevitably coupled with anxiety and a concomitant desire to regulate or control its consumption. Alcohol is not only universally subject to special rules and regulations, but also inspires powerful emotions. "Whether predominant feelings about [alcohol] are positive, negative, or ambivalent varies from culture to culture, but indifference is rare, and feelings are usually much stronger in connection with alcohol than with respect to other things."[4] Heath opens his landmark 2000 monograph on alcohol and culture with the following dedication, which wonderfully captures the Jekyll and Hyde nature of the ethanol molecule's impact on the human brain:

ETOH, alias Ethanol, Alcohol, C_2H_5OH

food, complement to food, and poison,
stimulant to appetite, and aid to digestion,
tonic, medicine, and harmful drug,
elixir, potion, or "a tool of the devil,"
energizer or soporific,
sacrament or abomination,
aphrodisiac or turn-off,
euphoriant, and depressant,
adjunct to sociability or means of retreat,
stimulant or relaxant,
tasty nectar or godawful stuff,
exculpatory, or aggravating, with respect to blame,
god's gift or a curse,
analgesic and anesthetic, disinhibitor or knock-out,
etc., etc.[5]

There are good reasons for seeing intoxication as a two-faced god. We have noted that alcohol, the king of intoxicants, has a "biphasal" effect: It initially functions as a stimulant, creating a positive, energizing state, but then transforms into a depressant. The neural fragments thrown off by this pharmacological hand grenade can also produce everything from a happy, extroverted back-slapper to an angry, belligerent sociopath. We dwelled at some length before on alcohol's positive individual and social functions. Passed over in relative silence has been the fact that, among mortality risk factors, alcohol consumption plays a starring role. The World Health Organization reported that in 2016 more than 3 million people had died as the result of misuse of alcohol.[6] The National Institute of Health estimates that alcohol is the third-highest preventable cause of death after smoking and lack of exercise.[7] No account of the role that alcohol has played in bringing suspicious apes together into creative, cultural, and communal civilizations is complete without also considering the dark side of Dionysus.

THE PUZZLE OF ALCOHOLISM

Alcoholism is an ancient human scourge, and probably the most crippling downside of alcohol in terms of the sheer quantity of damage it causes is to the alcoholics themselves, the people around them, and society at large. In one ancient Egyptian letter, written by a teacher to his former student,

> the teacher writes that he hears that his former student is forsaking his studies and is wandering from tavern to tavern. He smells of beer so much that men are frightened away from him, he is like a broken oar, which cannot steer a steady course; he is like a temple without a god, like a home without bread. The teacher ends by hoping that the student will understand that wine is an abomination and that he will abjure drink.[8]

This Egyptian student sounds like a classic sufferer of alcohol use disorder (AUD), the standard medical term for alcoholism. The anguish and helplessness felt by his former teacher would be familiar to anyone having to cope with a friend or loved one fighting, and losing, a battle against this disease. Today, global estimates of alcoholism rates range from 1.5 to over 5 percent of the general population, with large variations by country (Figure 5.1).

Share of population with alcohol use disorders, 2017
Alcohol dependence is defined by the International Classification of Diseases as the presence of three or more indicators of dependence for at least a month within the previous year. This is given as the age-standardized prevalence which assumes a constant age structure allowing for comparison by sex, country and through time.

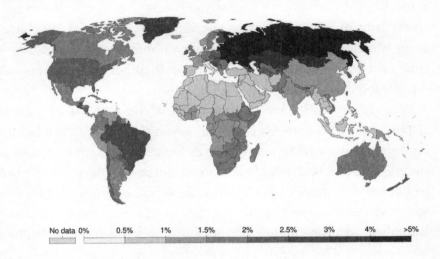

No data 0%   0.5%   1%   1.5%   2%   2.5%   3%   4%   >5%

Figure 5.1. Global Alcoholism Rates (from Our World in Data; ourworldindata.org/
alcohol-consumption)

In the U.S., 15.1 million adults are reported to suffer from various degrees of alcoholism, leading to 88,000 annual alcohol-related deaths and an estimated cost of $249 billion to the economy. As many as 10 percent of children live in households where at least one parent has a problem with alcohol abuse. Beyond the obvious costs and suffering caused by serious alcoholism, less dramatic but more widespread forms can have a pervasive, negative impact on our well-being. In one survey of American adults, almost 30 percent of respondents (and 36 percent of men) reported having experienced "mild" forms of alcohol use disorder, characterized, for instance, by repeated incidents of drinking more than intended, or by being unable to cut down on alcohol consumption despite a desire to do so.[9] An even more subtle danger is that drinkers can become dependent on alcohol not as a means for getting into an altered state, but merely to maintain a baseline state of happiness. Some addiction researchers believe that regular alcohol users neurologically adapt to it in such a way as to become dependent upon continued consumption merely to feel "normal." Habitual drug use can create a situation where homeostasis, the ability to maintain physiological stability of mood or affect, becomes reset at a pathological or harmful point.[10]

From an evolutionary perspective, our vulnerability to alcohol dependence and abuse is puzzling. Given that alcohol has been a common feature in human cultures for millennia, and that most people are able to drink alcohol in moderation, why does alcoholism exist? A tendency toward alcoholism is strongly heritable, with some scholars estimating the genetic contribution toward the likelihood of an individual developing the disorder to be as high as 60 percent.[11] The specific genes involved have not yet been clearly identified, but likely candidates are those that code for dopamine receptors, especially because people with alcohol addiction often show susceptibility to other types of addictions. It seems that those prone to alcohol abuse experience both enhanced reinforcement of alcohol's early, euphoric effects and a dampened sensitivity to the punishing effects on the downward curve of BAC.[12] Another line of research into alcoholism, led by addiction researcher Markus Heilig, has focused on genes associated with the function of the neurotransmitter GABA in the amygdala, a center of emotional arousal and fear processing. In both alcohol-dependent rats and humans, GABA activity in the amygdala is unusually low, which suggests that a genetically impaired ability to deal with negative arousal or stress may contribute to alcoholism.[13]

Whatever the underlying cause, it remains the case that up to 15 percent of the human population might be vulnerable to severe alcoholism, although not all of these individuals will actually become alcoholics. Given how long humans and alcohol have coexisted, this is genuinely baffling. Alcoholism is profoundly damaging and maladaptive. Why haven't the genes that give rise to it been driven out of the human gene pool? One would think that, in any part of the world where there is access to alcohol, there would be strong selection pressure against them. Another part of the puzzle is the wide cross-cultural variation in rates of alcoholism illustrated in Figure 5.1. Why is Italy near the bottom globally, despite widespread consumption of incredibly delicious alcoholic beverages (by both humans and lemurs), while Russia is at the top?

The answer to this cultural variation question might, in fact, be the key to unraveling the broader mystery of alcoholism. In terms of cultural variation within Europe, Italy is a classic example of what is sometimes called a "Southern" drinking culture.[14] In Southern Europe, alcohol—primarily wine, but also beer—is part of everyday life, so integrated into the cuisine that mealtime would be unthinkable without it. Children are introduced to moderate, healthy drinking habits early in life. In Italy, for instance, children from an early age are given glasses of wine mixed heavily with water, with the mix gradually becoming less and less diluted as they become older. People generally do not drink anywhere except at the lunch or dinner table, and drinking to the point of overt drunkenness is frowned upon. Distilled liquors are not unknown, but are generally consumed in very small quantities before or after the main meal, as an aperitif or aid to digestion. Overall alcohol consumption per capita tends to be quite high in these countries, but rates of alcoholism and alcohol-fueled disorder are low.

Although physically in Eastern Europe, Russia is a classic "Northern" drinking culture, as are, for instance, other Eastern European countries, Germany, the Netherlands, and the Scandinavian countries. Historically speaking, these cultures do not drink as much at home or with meals. Alcohol tends to be strictly forbidden to children, who come to think of it as an adult and somewhat taboo substance. Drinking as a primary activity, separate from mealtimes, is more common. Distilled spirits are frequently mixed in with the beer and/or wine, and may even replace them entirely. Northern drinking cultures drink less frequently, but are more prone to binge drinking. Public drunkenness is not uncommon, and in some cases it is seen as a badge of honor or manliness. Drinking alone, away from the meal table and outside of social contexts, is also not as stigmatized as in the South.

In Chapter One, we noted that the hijack and hangover theories see alcohol use as a pure vice, at least since we humans figured out agriculture, settled civilizations, and how to stock up on almost endless quantities of beer and wine. A taste for alcohol, especially the excessive and extremely harmful variety of craving found in alcoholics, should therefore have come under strong negative selection pressure over the past few thousand years. However, we also remarked upon the odd failure of genetic "solutions" to the "problem" of alcohol, such as the "Asian flushing" gene combination, to spread much beyond limited regions of Southeast Asia and the Middle East. This suggests that, at least over the past few thousand years, the benefits of being able to drink, and sometimes drink to excess, have historically outweighed the costs.

It is entirely possible, however, that this calculus has changed quite recently. Over the past few centuries—the blink of an eye on an evolutionary timescale—there have been two major innovations in the way that people produce and consume alcohol. The first was the advent of distilled liquors. The second was a set of changes in lifestyle and economics that has made drinking alone, or at least drinking completely outside the purview of social and ritual control, a real possibility for large portions of the human population. In the absence of appropriate countermeasures, these two innovations, distillation and isolation, might shift the cost-benefit calculus against alcohol use.

The degree of novel risk, however, is strongly moderated by cultural factors. Cultural norms have always functioned to mitigate the dangers of alcohol, and throughout history societies that have developed effective cultural norms to moderate and regulate alcohol have been better positioned to reap the benefits of alcohol consumption while minimizing the costs. It is therefore probably no accident that the lowest rates of alcoholism in modern Europe are found in Italy and Spain, and the highest in Northern and Eastern Europe. (Immigrants also bring their drinking cultures with them: Italian-Americans, for instance, have lower rates of alcoholism than the national average.)[15] Southern drinking cultures provide individuals with a genetic propensity toward alcoholism with effective safeguards against these novel problems, giving us a sense of how genes for alcoholism could have remained viable over the past few thousand years. The latter cultures expose potential alcoholics to the full force of distillation and isolation, making genes for alcoholism much more harmful than they historically have been. There are important lessons for us here when it comes to the question of how to tame or domesticate Dionysus, especially because the relatively defenseless Northern drinking attitude finds its apogee in the United States, and American culture in turn has set the standard for much of the world since the middle of the twentieth century.

Let us turn to the twin banes of distillation and isolation, and how they radically magnify the dangers that alcohol has always presented to human beings.

THE PROBLEM WITH LIQUOR: AN EVOLUTIONARY MISMATCH

For almost all of our long history with alcohol, the beverage in question has come in the form of a beer or wine. These beers and wines, in turn, have typically clocked in at around 2 to 4 percent ABV, or alcohol content by volume. Modern fermentation techniques, which rely on hyper-efficient and especially alcohol-tolerant yeasts, are capable of producing more powerful beers and wines, with contemporary beers averaging 4.5 percent ABV and wines 11.6 percent.[16] Any process of natural fermentation is inherently capped, however, by the alcohol tolerance of the yeast: At some point, even the most hardy yeast strains get killed off by their own by-product, effectively bringing fermentation to a halt. The furthest humans have been able to push this process is to around 16 percent ABV. This is the strength of a head-banging Australian Shiraz, which is made from sugar-rich, super-ripe grapes (courtesy of Australia's hot climates) and the most ethanol-tolerant yeasts. Anyone who has opened one of these bombs is familiar with the burst of alcoholic fumes that explodes out of a freshly uncorked bottle. In the Aussie-Kiwi wine rivalry, I come down firmly on the side of the Kiwis, much preferring the more elegant and refined cool-climate wines produced in New Zealand, but the Aussies deserve credit for brutally pushing the wine grape to its absolute alcoholic limit.

We should not be surprised that super-clever and alcohol-craving apes eventually managed to find a way around the inconvenient problem of wimpy yeast throwing in the towel at 16 percent ABV. One method for circumventing the natural limitations of yeast is called "fractional freezing." Fractional freezing relies on the fact that pure water congeals at 0°C (32°F, for you Americans out there), while pure ethanol must be quite a bit colder, -114°C

(-173.2°F). Mixtures of alcohol and water will freeze somewhere between those points, which is why engine coolant antifreezes initially employed methanol, a close chemical relative of ethanol. If you take some beer and put it out in *really* cold weather, chunks of ice will eventually form as the mixture cools down. Because of the nature of water-ethanol mixes, the bits that freeze are not pure water, but an ethanol-water combination, which means fractional freezing cannot completely separate ethanol from the water in which it is dissolved. But because the slurry left behind is slightly richer in alcohol than the chunks of ice pulled out, if this process is iterated several times it can produce a more potent alcoholic beverage. *Eisbock* beers produced in this fashion can reach 12 percent ABV. On the American frontier, the fruits of Johnny Appleseed's trees were often turned into applejack, a potent liquor made from freeze fractionating apple cider, which generally came in at around 20 percent ABV.

Prior to modern times, of course, this process was limited to regions of the world with very cold winters. To get an applejack of 20 percent ABV, for instance, you'd need to start with an already pretty potent cider and then cool it down to -10°C / -23.3°F. Moreover, fractional freezing is an inherently crude process. Because the water and ethanol remain always intermingled, the chunks of ice pulled out contain a greater and greater proportion of ethanol, so there are limits to how powerful the end product can be. Another significant problem is that the mixture left after the ice chunks are removed is rich not only in ethanol but plenty of other more nasty stuff, including other forms of alcohol and organic molecules that are either toxic or foul-tasting or both. Desperate American pioneers were willing to put up with these often noxious concoctions, but there is a good reason you no longer see freeze-fractionated liquors on contemporary restaurant drink lists.

Overall, then, freeze fractionation is a geographically constrained, inefficient, and crude process for boosting alcohol content. The real killer app, if you want to get super-wasted in a hurry, is distillation. Distillation is both elegant and simple, at least conceptually. Take your mixture of water and ethanol and *heat* it rather than cool it. Water and ethanol are both relatively volatile, which means they boil off well before the other chemical components of a beer or wine do. (This is why distilling water is a good way to purify it for drinking—boil some dirty water and the H_2O comes off as a vapor that can be collected, leaving germs and undesirable organic molecules behind.) Conveniently for those looking for a concentrated alcoholic bang, ethanol is even more volatile than water, boiling at 78.3°C (173°F) rather than water's 100°C (212°F). This means that if you heat a beer or wine, the ethanol will boil off first. If you can figure out some way to capture that alcoholic vapor and cool it back down into a liquid, voilà, you've got yourself some more or less pure alcohol. Break out the shot glasses.

The problem is that, in practice, distillation is fiendishly difficult to pull off. As Adam Rogers notes, distillation "requires the ability to boil a liquid and reliably collect the resulting vapors, which sounds simple. But to do it, you have to learn a lot of other skills first. You have to be able to control fire, work metal, heat things and cool them, make airtight, pressurized vessels."[17] You need to be able to precisely control the temperature of various liquids and vapors, and know when in the heating process the vapor being generated is ethanol instead of something else you might not want. In addition to being technically rather challenging, distillation is *dangerous*. Exploding home stills and scalding liquids were Prohibition-era America's equivalent of contemporary meth lab disasters.

Yet we are a determined and resourceful species of ape. The

principles of alcoholic distillation, as well as distillation as a water purification method, were described by Aristotle, and there are suggestions of distillation being practiced on an experimental scale in ancient China, India, Egypt, Mesopotamia, and Greece.[18] By the Middle Ages, we read of alcoholic stills in Persia and Tang Dynasty (618 to 907) China. The former gave us our English word "alcohol," from the Persian phrase for distilled ethanol, the *al'kohl'l* or "mascara" of wine.[19] The latter produced texts that begin speaking of *shaojiu* or "cooked/distilled wine," and banqueting cups from this period begin to shrink in size, perhaps reflecting a shift in elite circles from beers and wines to distilled liquors.[20] Distilled liquors did not really become widespread, however, until relatively recently, probably the thirteenth century in China and sixteenth to eighteenth century in Europe.

This fact is very important for the story we have told above. If alcohol has played a crucial role in helping to catalyze civilization, creativity, and human cooperation, for most of its 9,000-plus-year history it has been in the form of relatively weak beers and wines. If wine represents a jump in alcoholic content (11 percent ABV) with respect to some overripe grapes fallen from a vine (3 percent ABV), the brandy made from distilling that wine (40 to 60 percent ABV) is a quantum leap. The early Greeks were deeply concerned about the dangers of drinking undiluted wine, a barbarian practice they believed led inevitably to violence and chaos. They would have been absolutely horrified by the potential for chaos contained in a bottle of brandy.

Distilled liquors are not only wildly more potent than naturally fermented beverages, but they also preserve extremely well and are easy to package and ship. The historian Daniel Smail argues that a crucial marker of the beginning of what we think of as "modernity" was the moment when chemical intoxicants previously confined to various parts of the world—caffeine in Africa, nicotine in the Americas, opium in Central Asia—"fell together into a new [global] framework."[21] One prominent feature of this new global network was the trading of rum, gin, and other distilled spirits, which remain potent and drinkable for decades and can be easily shipped to every corner of the globe. The advent of distillation therefore radically changed the scope and reach of alcohol consumption. Distillation is what makes it possible for almost anyone, anywhere in the industrialized world, to walk into a corner store and emerge a few minutes later, and only a few dollars lighter, with a truly insane quantity of alcohol tucked into a small brown paper bag. A couple bottles of vodka contain a dose of ethanol equivalent to an entire cartload of pre-modern beer. The availability of such concentrated intoxicants is entirely unprecedented in our evolutionary history, and not a good development for potential alcoholics.

Liquor also seriously distorts social drinking, because it gets you very drunk, very fast. What the Germans call a *Schwips,* a pleasant social buzz, characterizes one's mental state from the first few sips of a drink through about .08 percent BAC, the point at which most jurisdictions consider you to be legally drunk. People socially drinking beer or wine, especially in the context of a meal, rarely get beyond .08 percent or so. This is good, because things quickly go downhill after that. By .10 percent BAC one is quite drunk, and .30 percent is about as drunk as most people generally get, even in a wild night of debauchery. This is the point at which alcohol's depressant effects start to swamp out everything else, and it is characterized by the slurred speech and difficulty walking of a drinker who really needs to go home. Most people pass out by .40 percent BAC, which is actually a blessing, because exceeding that level can cause such intense physiological depression that breathing and heart function can stop.

It is very difficult to pass out from drinking beer or wine; it is nearly impossible to kill oneself. Once distilled liquors are in the mix, however, all bets are off. The rapidity with which one can get dangerously drunk on gin or vodka is frightening. Unlike beer or wine, the speed and force with which it hits our nervous systems makes liquor difficult to integrate harmoniously into social events or meals. People doing vodka shots blow right past the sweet spot of the .08 percent BAC social *Schwips* on an express train from fully sober to slurred speech and wild disorientation. For those with alcoholic tendencies, distilled liquors provide the quickest and surest route to dependence.

Despite ancient warnings about the dangers of drunkenness and disorder, our only genuine mass epidemics linked to alcohol have been relatively recent, because they have been fueled by liquor. In eighteenth-century Britain, for instance, the sudden availability of large quantities of cheap spirits led to a "gin craze" that terrorized London and led to a wild spike in crime, prostitution, poverty, child abuse, and premature death.[22] A massive decrease in life expectancy occurred in Russia after the demise of the Soviet Union in 1991. When full market reforms hit and the state alcohol monopoly was abolished, the price of vodka plummeted relative to other goods. Between 1992 and 1994, life expectancy in Russia fell 3.3 years for women and an incredible *6.1* years for men, with later research suggesting this rise in mortality was driven by a massive increase in vodka consumption.[23]

The many functional benefits of alcohol notwithstanding, distillation radically increases its danger to both individuals and societies. And it is a novel threat. People sometimes have trouble thinking on evolutionary timescales, and 1500 CE might seem like a really long time ago. So to provide a visual sense of how recent a development distillation is, Figure 5.2 indicates when, in the long history of our primate lineage's adaptation to alcohol, distillation arrives on the scene.

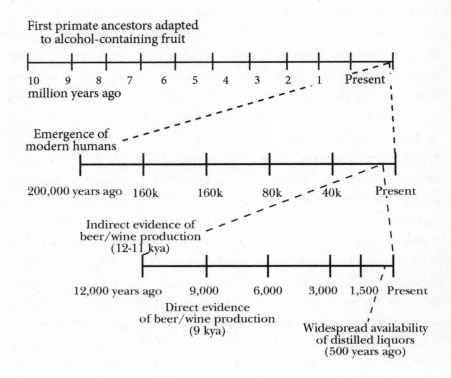

First primate ancestors adapted
to alcohol-containing fruit

| | | | | | | | | | |
10   9   8   7   6   5   4   3   2   1   Present
million years ago

Emergence of
modern humans

200,000 years ago   160k   160k   80k   40k   Present

Indirect evidence of
beer/wine production
(12-11 kya)

12,000 years ago   9,000   6,000   3,000   1,500 Present

Direct evidence
of beer/wine production
(9 kya)

Widespread availability
of distilled liquors
(500 years ago)

Figure 5.2. Timeline indicating when our first primate ancestors adapted to alcohol-containing fruit (10 million years ago), the emergence of modern humans (200,000 years ago), indirect evidence of beer and wine production (12,000 years ago), direct evidence of beer and wine production (9,000 years ago), and widespread availability of distilled liquors (500 years ago).

While the sixteenth century might seem like ancient history, it was basically yesterday in evolutionary terms.

Distillation is not only a novel danger, it is one that also facilitates a companion peril: drinking outside of social contexts. Let us turn now to considering not only the sheer alcoholic punch contained in the corner store bag of liquor, but also the risk of being allowed to simply walk home alone with it.

ISOLATION: THE DANGER OF DRINKING ALONE

If you've ever complained about how long it takes to get a post-work drink at a crowded pub on a Friday evening, you should be thankful that you don't live in ancient China. An early Chinese ritual text describes the beginning of the traditional wine-drinking ritual as follows:

> The host and the guest salute each other three times. When they reach the steps, they concede to each other three times. Then the host ascends. The guest also ascends. The host stands under the lintel, faces north and salutes twice. The guest ascends from the west of the steps, stands under the lintel, faces north and returns the salutation. The host sits down and takes the [wine] cup from the tray and descends to wash. The guest follows the host. The host sits down again and pronounces his words of courtesy, and the guest replies.[24]

As the sinologist Poo Mu-chou notes, "The ceremony itself was designed to celebrate the friendship between the participants with the help of wine drinking, although the actual drinking of wine comes at the end of this long procedure." Even after this preliminary ballet of salutations and courtesies and ritual washing of the cups is finally over, the ancient Chinese tippler is still not free to consume at will: One does not drink unless a formal toast is made, and who has the right or responsibility to make a toast is also strictly dictated by ritual.

This means that the host of a traditional wine-drinking banquet, by adjusting the frequency of toasts, can thereby effectively regulate his guests' level of inebriation. If conversation and good cheer are lagging, toast frequency goes up; if things begin to get out of hand, it's time now to focus on the vegetable dish. For the early

Chinese, the ritualized wine banquet served as a basic metaphor for governing in general because of the elegance with which it tamed a potential source of social chaos—alcohol—through ritual order, restraining and regulating its use.[25] As the ancient Chinese historian Sima Qian explains:

Feeding pigs and making wine [for a banquet] do not in themselves lead to disaster. But lawsuits grow increasingly frequent, and it is the pouring of wine that brings about the disaster. Thus the former kings made the drinking rituals on this account. The ritual of offering wine, making the hundred obeisances of guest and host, allow people to drink all day but never become drunk. This was the measure taken by the former kings to ward off disasters incurred by wine. Wine and feasting are used to create shared happiness; music is used to demonstrate virtue; and the rites are used to stave off excess.[26]

Cultural strategies for "staving off excess" when it comes to drinking are as widespread as alcohol itself. From ancient Sumer and Egypt to ancient Greece, Rome, and China, written or pictorial representations of drinking always depict the consumption as being both social and socially regulated. The host of the Greek symposium, for instance, not only controlled the timing and ordering of the toasts, but also the water to wine ratio being served, adjusting it up or down as deemed necessary.

This sort of social regulation is also a feature of many contemporary cultures. The Tohono O'odham people living in the Sonoran Desert home-brew an alcoholic beverage made from fermented cactus juice, but "no family may drink its own liquor lest the house burn down, [although] they may drink at other houses"—a taboo that effectively makes consumption a public act, and therefore one subject to social control.[27] In traditional house-holds in Georgia, the Head of Table (*tamada*) controls alcohol consumption very much like a traditional Chinese banquet master or host of a Greek symposium, judiciously spacing out the toasts and also holding the authority to bring the proceedings to an end when it is clear that everyone has had enough.[28] In Japan, certain Shinto rituals demand inebriation on the part of participants, but levels of drunkenness are carefully monitored by the group, and individuals who have been too enthusiastic in their ritual drinking are gently escorted home.[29]

This strategy is effective because, in most societies and for most of human history, the consumption of chemical intoxicants, especially alcohol, has been a fundamentally *social* act. In the vast majority of societies, no one drinks alone. Imbibing is communal and highly regulated by both formal and informal rituals. As Dwight Heath concludes in his survey of cross-cultural uses of alcohol, "Solitary drinking, often viewed as a crucial symptom of problem drinking, is virtually unknown in most societies."[30] To the extent that solitary indulgence in intoxicants occurs, it is widely condemned or viewed with suspicion. Among mestizo communities in the highlands of Peru, reports the anthropologist Paul Doughty, "Drinking is a social act and is part of virtually every social gathering. Solitary drinkers are considered deviant and at best are looked upon as unfortunate individuals, or at worst, as unfriendly or 'cold' (*seco*)."[31] In Oceania, "'drinking kava alone' is an idiom for witchcraft"—anyone indulging alone must be up to no good.[32] Even in the United States, home to perhaps the most highly individualized and fragmented drinking culture in the world, drinking alone carries a certain stigma. It is no accident that the 1985 hit "I Drink Alone" portrays an out-of-control, solitary alcoholic binging on distilled liquors, and comes from the same singer, George Thorogood, who gave us "Bad to the Bone."

In this respect, it is revealing that kava, which is smoothly integrated into the social lives of cultures that have traditionally used it, has taken on the role of a dangerous and seriously abused drug in more recent times when exported to other regions. For instance, consumption among Australian indigenous populations, who have no history of kava use, ranges to as high as fifty times the amount drunk in the Pacific Island cultures where kava was originally domesticated. This has led to enormous individual and social problems. Researchers attribute this disparity to kava having been torn out of its traditional ritual and social context, removing important constraints upon individual consumption.[33]

It is clear how formalized drinking rituals or ceremonies regulate individual intoxicant use. What is perhaps less obvious is how even completely informal get-togethers, or indeed any sort of public drinking, inherently involve a degree of social monitoring and control. An ethnographer in Norway who studied a group of twenty-somethings notes that even at the fairly chaotic drinking parties where young Norwegians throw back shocking amounts of alcohol, there is an at least implicit "emphasis on healthy drinking as being underpinned by ideas of collectivity and by group responsibility for individual consumption." When one person in the friend group began drinking alone at home before the parties started, this was viewed as a problematic sign and a cause for intervention.[34] During the drinking parties, it was considered bad form to throw away or recycle empty bottles, which instead accumulated in front of each drinker, giving everyone an instant and accurate sense of how much each person had drunk. Friends who accumulated too many empty bottles too fast were observed to unconsciously regulate themselves, slowing down their consumption to better track the group.

This phenomenon, sometimes referred to as "matched drinking," has been widely observed in cultural contexts around the world and can be replicated in the lab.[35] However, given that individuals sometimes "match up"—i.e., increase their consumption to accord with their peers—this can lead to more, rather than less, drinking. And it is true that in pathological forms, like fraternity hazing rituals, this can lead to terrible outcomes. Cultures designed exclusively by feral young men ripped apart from their families and communities, however, are fairly rare outside of the university context or *Lord of the Flies*. Most cultures set reasonable limits on alcohol consumption, and the crucial point is that individuals' levels of inebriation are brought under social control when they drink in groups, even quite informal ones.

Laboratory research also shows that people in social drinking conditions report increased levels of "positive mood, elation, and friendliness," whereas subjects required to drink in isolation report higher levels of depression, sadness, and negative emotion.[36] Drinking in a group also seems to protect against the alcohol-induced increase in risky behavior discussed below—collective group opinion appears able to compensate for the biases resulting from individual, alcohol-induced cognitive myopia.[37] As one research team notes, the group monitoring that occurs in social drinking means that "drinkers may be relatively protected while with a group. They may be able to 'watch one another's back.' In contrast, a solitary drinker may be in a relatively more unpredictable and vulnerable state when alone."[38] Again, this process can go seriously wrong in unhealthy frat or other hazing cultures, but generally serves to reduce and normalize alcohol consumption.

In our modern world, drinking too often occurs in a social vacuum.[39] This is especially the case in suburban communities, where people commute long distances from home to work, trapped in their individual box on wheels while in between. Suburbanites typically also lack a social drinking venue within easy walking distance, where they might continue conversations begun earlier in the day or unwind with other regulars between work and dinner. Drinking increasingly occurs only in private homes, outside social control or observation. Knocking back a string of high-alcohol beers or vodka and tonics in front of the TV, even with one's family around, is a radical departure from traditional drinking practices centered on communal meals and ritually paced toasting. It instead calls to mind the bottomless alcohol feeding tubes provided to overcrowded rats in those alcohol and stress experiments. Individualized, on-demand delivery of strong booze is as unnatural for humans as it is for rats.

DISTILLATION AND ISOLATION: THE TWIN BANES OF MODERNITY

The widespread availability of distilled liquors and a rise in solitary drinking are relatively recent developments, and may fundamentally change alcohol's balance on the razor's edge between usefulness and harm. Large-scale epidemics of alcohol abuse are invariably driven by one or both of these banes of modernity, with distilled liquor becoming particularly harmful when the availability of spirits coincides with a breakdown of social order or ritual regulations. We can see both forces at work in the Russian vodka epidemic that followed the breakup of the Soviet Union, and also when it comes to the problem of alcoholism among Native American populations. The historian Rebecca Earle has written extensively on the prejudice of the "drunken Indian" in Spanish colonial America, a trope that was exaggerated and exploited by missionaries and colonizers to justify their oppression and disappropriation of native populations. Nonetheless, as Earle notes, alcohol *did* become more problematic for indigenous communities in the post-colonial era, and our twin banes seemed to be the primary causes. This era was characterized by a breakdown in religious rituals, the social regulatory mechanisms that previously allowed various South American cultures to safely incorporate alcoholic beverages, like *chicha* or *pulque,* into their daily lives. It also featured the introduction of distilled liquors, which pack an alcoholic punch far more powerful than indigenous fermented beverages.[40]

In this context, it should not surprise us that Russia, which leads the world in rates of alcoholism, remains characterized by a somewhat fragmented social order combined with an almost exclusive taste for distilled liquor. A glance at Figure 5.1 will show that Americans are also no slouches when it comes to high rates of alcoholism. This can probably be attributed, at least in part, to the extreme individualism and dispersed suburban living styles that characterize the States, at least as compared to European countries. America is one of the few places in the industrialized world where having a local pub or café is rare, and where drive-through stores allow an individual to acquire cigarettes, firearms, Slim Jims, and enough alcohol to paralyze an elephant without having to leave the comfort of her SUV. This is a historically unprecedented lifestyle, and one for which we may not be evolutionarily well equipped, genetically or culturally.

We noted above that alcoholism is strongly heritable, and raised the question of why the genes predisposing one to alcoholism have remained in the human gene pool. One possibility is that, prior to the advent of distillation and unregulated private drinking, the dangers of alcoholism were outweighed by the individual and social benefits of a taste for alcohol, but that this calculus has since changed. In a world awash with powerful distilled spirits, and where drinking increasingly takes place in the privacy of one's own home, alcohol may indeed be more dangerous than helpful. It is quite possible that distilled liquor is such a novel threat that genetic evolution simply hasn't had time to catch up.

In Chapter One I dismissed "hangover" theories that argue that alcohol may have been adaptive in our distant evolutionary past, but became maladaptive as soon as humans invented agriculture and could produce beer and wine in quantity. But in identifying the novel dangers of distillation and isolation, I am essentially opening the door to a modified hangover theory—one in which the adaptive buzz provided by alcohol turned into a painful headache much more recently, in the past few hundred years or so. If this is true, one prediction would be that we will start seeing the genes that contribute to the "Asian flushing syndrome," and which are protective against alcoholism, begin to spread beyond their current geographically constricted range. In this context, it is worth noting that East Asia not only has the greatest concentration of these genes, but also had a 300- to 400-year head start on the rest of the world when it comes to widespread distillation.

A rapid and dramatic change in our adaptive environment would also clearly be a situation where cultural evolution can leap into action. The challenges presented by distillation and isolation may require us to seriously rework our cultural methods for dealing with the world's most popular drug. In this respect, Southern drinking cultures, which seem to have dealt relatively well with both of these threats, may provide a model. We will outline some of the useful features of this drinking culture at the end of this chapter.

Before that, however, it is time to focus our discussion on some other ways, short of out-and-out alcoholism, in which things can go seriously off the rails when intoxication is poorly regulated. The topic of the costs that alcohol imposes on individuals and societies has come up previously in the course of arguing against hijack or hangover theories. So far, these costs have been mentioned only in passing. In weighing how we should feel about the adaptive value of alcohol in the modern world, it is important to explore them in more detail, as well as to elaborate on the dark underbellies of some of the adaptive functions discussed in Chapters Three and Four.

DRUNK DRIVING, BAR FIGHTS, AND VENEREAL DISEASE

When discussing the costs of alcohol consumption, the phrases "contributions to mortality," "alcohol-related deaths," and "alcohol-related harms" frequently pop up. These wonderfully vague health policy terms refer to outcomes that go far beyond ruined livers. Alcohol certainly does hurt your body, especially if taken in excess, but alcohol-related harms "include a wide range of negative consequences such as loss of productivity, violence, injuries, academic failures, unintended pregnancy, sexually transmitted diseases, cardiovascular diseases, cancer, etc."[41] Perhaps the most obvious negative behavioral consequence of alcohol misuse is drunk driving,[42] but the WHO links alcohol to a wide variety of deaths, including those caused by liver damage, cancer, self-harm, industrial accidents, poisonings, drownings, falls, and the rather capacious category of "other unintentional injuries" (Table 5.1).

Alcohol-attributable fractions (AAFs) for selected causes of death, disease, and injury, 2016

Cause	All Global Deaths	All Global Disability-Adjusted Life Years
Alcohol use disorders	100%	100%
Cirrhosis of the liver	48%	49%
Other pharynx	31%	31%
Road injury	27%	27%
Lip and oral cavity	26%	26%
Pancreatitis	26%	28%

Cause	All Global Deaths	All Global Disability-Adjusted Life Years
Larynx cancer	22%	22%
Tuberculosis	20%	21%
Self-harm	18%	19%
Interpersonal violence	18%	18%
Esophagus cancer	17%	17%
Exposure to mechanical forces	14%	15%
Other unintentional injuries	14%	13%
Epilepsy	13%	10%
Poisonings	12%	10%
Drowning	12%	10%
Falls	11%	15%
Colon and rectum cancers	11%	11%
Fire, heat, and hot substances	11%	11%
Liver cancer	10%	10%
Hemorrhagic stroke	9%	10%
Hypertensive heart disease	7%	8%
Cardiomyopathy, myocarditis, endocarditis	7%	8%
Breast cancer	5%	5%
Lower respiratory infections	3%	2%
HIV/AIDS	3%	3%
Ischemic heart disease	3%	2%
Ischemic stroke	-1%	-1%
Diabetes mellitus	-2%	-2%

Note: For ischemic stroke and diabetes mellitus, the AAFs were negative, meaning that, overall, alcohol consumption has a beneficial effect on these diseases.

Table 5.1. The "alcohol-attributable fractions" (AAF) for selected causes of death, disease, and injury, worldwide, in 2016.[43]

The brutal health consequences caused by heavy drinking, as well as the misery and suffering inflicted on the world by drunk drivers, have been widely discussed, and need not be elaborated upon here.[44] These tragic costs alone weigh heavily on the negative side of the scale when we evaluate the role of alcohol in human society. Of the many other negative consequences attributable to drink, it is worth focusing here on two particularly salient ones: aggression and generic risk-taking.

Alcohol is the only drug, besides pure stimulants like meth, that is known to increase physical aggression and violence.[45] Cannabis, kava, MDMA, and psychedelics all produce either mellow or introverted highs. Alcohol's stimulating effect, when combined with cognitive myopia and loss of executive function, can induce aggressive or violent behavior, especially in people with already low levels of cognitive control.[46] As one writer notes, cultures have always been rightly leery about mixing alcohol and crowds, especially in situations likely to elicit strong emotions, like sporting events. One Greek classicist describes an inscription at the stadium in Delphi, dating to 5 BCE, that forbids spectators from bringing wine into the arena, adding that similar warnings are still posted outside the football stadiums at Harvard and Southern Methodist Universities.[47] As any European football fan knows, adding alcohol to the mix of large crowds, strong emotions, and intense team rivalries is a recipe for widespread violence and hooliganism.

In addition to lowering the barrier to aggression, alcohol's handicapping of cognitive control increases general risk-taking. In one study,[48] experimenters had four groups of subjects—a placebo condition, and three groups that were brought to increasing levels of intoxication with spiked ginger ale—play a series of games measuring their ability to intuitively weigh a payoff option that returned safe and small rewards against a flashy, but risky and ultimately more costly, alternative. Given a small pot of money ($6.00) to start with at the beginning of each trial, they were repeatedly asked to choose between two payoff options, labeled "C" and "A." C was the safe option: clicking on it invariably resulted in a small, reliable gain of $0.01. Sitting in front of a computer repeatedly clicking "C" was a rather dull experience, but gave a predictable and more rewarding outcome in terms of real money the subjects could take home with them. The risky option, A, delivered more excitement, generating random payoffs or costs ranging from $0.25 to a full dollar, but was designed to be more costly overall. Choosing A would provide a more exciting ride, but leave the subject worse off financially by the end of the experiment.

Figure 5.4 shows the change in risky versus safe responses between the initial, pre-dosing session and session 2, where subjects in the alcohol conditions reached their peak BAC.

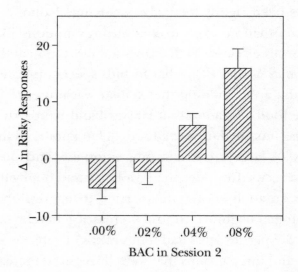

Figure 5.4. Change in risky versus safe responses from session 1 (pre-dosing) to session 2 (peak alcohol effect) across four conditions, from placebo (.00) to .08 percent BAC. The change was calculated by subtracting a number of risky choices in session 2 minus the same number in session 1.[49]

Subjects in the placebo condition and those given weak doses of alcohol (brought to just under .02 percent BAC) quickly learned to avoid the risky option, reducing the number of times they chose it and ending up with a better payout than those who had been brought to just above .04 percent or .08 percent BAC. The inability to resist the exciting but costly option increased dramatically with BAC level. It should not surprise us that the risk task used by these experimenters is a variation of one used to study patients with PFC damage, and that this same pattern of being attracted to immediately appealing but ultimately less rewarding options is found in this population.

The relative insensitivity to negative feedback or long-term consequences induced by alcohol can lead to many dangerous behaviors, ranging from drunk driving to unsafe sex. The latter leads us to the topic of the dark side of alcohol's aphrodisiacal qualities.

BEER GOGGLES AND VIOLENCE AGAINST WOMEN

One of the more disturbing events in the book of *Genesis* (and that's saying something) is when the daughters of Lot get him almost insensibly drunk in order to seduce and become impregnated by him (*Genesis* 19:33). This is only the first in a long line of literary portrayals of alcohol as date rape drug.[50] Chapter Four discussed the role of alcohol in enhancing intimacy, allowing strangers to open up to one another and romantic couples to get beyond awkwardness or unhelpful inhibitions. Here we must balance that account by noting that alcohol, especially in excess, can also dangerously distort and harm romantic and sexual behavior. "The use of alcohol to relax sexual inhibitions or to enhance romantic and sexual feelings is normally not a problem; it is a delight," the authors of a review article on the effects of alcohol on sexual behavior concede. "Nevertheless, it should be noted that alcohol has also been implicated in numerous problematic sexual outcomes, including unwanted pregnancy, sexual dysfunction, sexual assault, and sexually transmitted diseases (including HIV/AIDS)."[51]

The term "beer goggles" is sometimes used in a gently humorous way to refer to alcohol's enhancement of the perceived attractiveness of others, an effect we discussed in Chapter Four. It can also, however, lead to potentially very harmful behavior. Directly examining the effect of beer goggles on human sexual courtship is ethically challenging to pull off in the lab, but we've noted that fruit flies provide a fairly good model for humans when it comes to the cognitive and behavioral effects of alcohol. Studies have shown that intoxicated male fruit flies sexually pursue everything in their environment, including other male fruit flies (not a typical behavior), displaying both enhanced sexual arousal and reduced discrimination.[52] Under the influence, female fruit flies similarly display reduced choosiness and weakened mate preferences.[53] What we might call the "beer mask" phenomenon, the fact that intoxicated people feel themselves to be more attractive also clearly has a dark underbelly. Studies have demonstrated that intoxicated, heterosexual men are more likely to misread female behavior as sexually suggestive.[54] Significantly, there was a clear specificity to this bias: Men in one study showed reduced ability to distinguish between generic friendliness and sexual interest while remaining capable of accurately processing other relevant cues, such as how provocatively the women were dressed.[55]

One recent study also showed that alcohol flips men into a more objectifying stance toward women, compared to sober controls, causing them to focus more on women's bodies than their faces.[56] The authors of this study, who cite lyrics from the song "Blame It (On the Alcohol)" as an epigraph, observe that "adopting

objectifying gazes toward women leads perceivers to dehuman-
ize women, potentially laying the foundation for many negative
consequences such as sexual violence and workplace gender discrimination."[57] A probably related, and particularly disturbing,
finding comes from studies where men are shown pornographic
clips depicting (fictionally) either consensual sex or rape and
have their physiological arousal measured. Sober male subjects are
more aroused by portrayals of consensual sex, whereas intoxicated
men are aroused by both. The self-reported readiness of men to
behave in ways that would constitute rape also increases with BAC.
It should therefore not surprise us that when sexual assault occurs
it is very common to find alcohol in the mix. Studies of convicted
rapists find that anywhere from 40 to 63 percent were intoxicated
when they offended.[58] Heavy use of alcohol also plays a central
and odious role in sexual assaults on college campuses, and in the
wider world is predictive of partner abuse.[59]

It is easy to conclude from this that alcohol, men, and women
should never mix. It is almost certainly the case that throwing
large groups of already prefrontally challenged young people into
house or frat parties with music, dancing, and massive quantities
of distilled liquor served in opaque red plastic cups is a recipe for
trouble. Analyses of police records find that in up to 72 percent of
the cases of sexual assault, either the offender, the victim, or both
were drunk, and alcohol is an important factor in acquaintance-
initiated sexual assault.[60] Ramping up sexual desire, reducing
inhibitions, impairing assessment of risk, and inducing debilitat-
ing cognitive myopia is a potentially toxic and dangerous cocktail.
But our takeaway from acknowledging this particular dark side of
Dionysus is complicated. The Mr. Hyde aspects documented here
coexist with the more positive Dr. Jekyll-ish functions documented
in the previous chapters, such as fostering group bonding and
enhancing trust. It is also arguably the case that the link between

alcohol and violence against women is driven by patriarchal or misogynist social norms rather than the ethanol molecule itself. Alcohol disinhibits, but it doesn't itself create the behavioral tendencies that are then unleashed. In any case, when it comes to alcohol and sex, we couldn't have a better example of alcohol's Janus-faced nature, its status as Heath's "elixir, potion, or 'tool of the devil.' "

Let's now turn to the less immediately devastating but nonetheless harmful downside of another of alcohol's sociality-enhancing function: how bonding and connection can lead to the perpetuation of cliques and insider groups.

OUTSIDERS AND TEETOTALERS NOT WELCOME: REINFORCING THE OLD BOYS' CLUBS

As we have seen, scholars have argued that the more or less obligatory sessions of after-work boozing inflicted on (almost entirely male) Japanese salarymen may have played an important role in defusing hierarchy and allowing otherwise sclerotic Japanese companies to continue to innovate in the 1980s and '90s. One observer waxes almost lyrical on the subject. In contrast to the dry world of the daytime office, the scene after hours is:

a more intimate, twilight, "happy hour" world of darkness, lit by neon signs and the painted smiles of more and less attractive hostesses whose job is to murmur sweet nothings and keep their clients' glasses full, while praising their—often better unremarked—renderings of popular karaoke songs. This world is known, somewhat poetically, as "the water trade" (*mizu shōbai*), and is said to provide the informal heart that enables formal business negotiations to be concluded.[61]

This rosy account rather leaves out, unfortunately, the unwelcome comments and touching endured by these "more and less attractive hostesses," or how the entire system serves to reinforce oppressive gender norms, a culture of sexual harassment, and the abysmally low participation rate of women in the Japanese workforce.

Chinese business banquets can be similarly horrifying experiences for women. An insightful, humorous op-ed by the author Yan Ge concerns her misadventures navigating the alcohol-soaked professional circles of Chengdu, her home city.[62] Thinking she had been invited to a casual dinner, but wandering instead into a full-on formal banquet, she plopped down next to what appeared to be the male host and realized with horror that she might be in the spot traditionally reserved for "the girl." As she explains, this refers to "the young woman installed to entertain the important middle-aged man. Those in this seat can expect to receive: an absurd amount of secondhand smoke; a number of judgmental looks from men and women around the table; inexhaustible bai-jiu [sorghum liquor] refills; and, occasionally, a squeeze on the shoulder or a hand on the back." Yan describes the physiological toll imposed by the marathon drinking that is the main activity at these events, and how it tends to reinforce traditional gender roles and be particularly excruciating for young women, who are expected to flirt and entertain their senior male colleagues. Her conclusion about the point of the Chinese business banquet is spot on:

The ultimate purpose of a banquet is to get its diners drunk. Only in this way can we connect and become friends, squeeze each other's shoulders and make dirty jokes. When it goes wrong, it can be ugly: Fights can break out; women might be abused for sport. But when it goes right, mistakes are forgiven; the diners perspire, devour, quaff and sing together, and then, only then, will business be done.

This nicely encompasses both the positive functions described in Chapters Three and Four and the worries that concern us here. Seriously impairing the PFC can clear the way for trust and forgiveness and generosity; it can also open the floodgates to hostility and misogyny.

While the UBC pub where our large research consortium got its start provided a somewhat more salubrious environment than a Japanese hostess bar or typical Chinese banquet, it is worth noting that the attendees of what came to be known as our Friday afternoon "centre meetings" were almost exclusively men. We were able to gently downregulate our PFCs and brainstorm creative ideas over a few pints because our wives agreed to do childcare pickup on Fridays, and to not be overly annoyed with us if we showed up for dinner a bit late or buzzed or both. Female colleagues were welcome, indeed encouraged, to join, and occasionally did. But it was usually about as male-dominated as the Japanese water trade. This, in turn, is likely due to the fact that unfair gender norms make it seem, consciously or not, more acceptable for men rather than women to miss daycare pickup because they are off at the pub. Our choice of timing and venue wasn't intentionally designed to create an atmosphere hostile to women, but may very well have done so inadvertently.

It is not immediately clear what the right way to think about this is. Given the demonstrable payoffs of this sort of alcohol-lubricated brainstorming, it seems counterproductive to declare that it should never happen. And yet there are obvious dangers of exclusion and inequity if this form of socializing occurs in venues that are unwelcoming to women, or even perceived to be so. It is revealing that the prohibition and patents study described in Chapter Four found that the imposition of prohibition reduced patent applications by *men*, but not women. The study's author sees this as an additional bit of evidence showing that the decline resulted from lack of communal drinking, because, in America in the 1930s, it was primarily men who gathered in the saloon to drink and exchange ideas. It also, however, points to one of the fundamental worries we should have about the social function of the pub or whiskey room: These are venues traditionally dominated by men. As long as the sharing of childcare duties between men and women, which in most industrialized societies has improved since the 1950s but still falls far short of 50-50, remains unequal, the ability to join in, and therefore benefit from, this kind of networking disadvantages women. Skewed gender ratios make it more likely that these events will turn into enablers of sexual harassment and assault, which in turn can serve to further reinforce the gender skew.

We mentioned in the last chapter the column bemoaning the demise of the Gay Hussar restaurant, a Soho drinking venue where politicians, writers, and activists would mix. That piece does an excellent job of highlighting what is lost as the influence of booze is gradually drained out of public life. When it begins, however, with "for political journalists of a certain age it is impossible to read about the imminent closure of the Gay Hussar...without a flood of nostalgia," one cannot help but assume that the vast majority, if not all, of these "political journalists of a certain age" are white men. Although intoxication functions to bind together non-related individuals, the bonds it creates can be deeply tribal. Disinhibition leads to honesty and talkativeness. There is something to be said, however, for people with prejudices or sexist views keeping their mouths shut. Socializing under the watchful eye of a fully functioning PFC may be a bit dull, but it does leave one's abstract reasoning abilities intact, potentially allowing people to rely less on implicit biases and more on objective qualities and shared, abstract goals. All in all, there might be something to be said for having a "professional political class," despite the author's complaints.

Alcohol-fueled socializing also puts those who do not drink, for whatever reason, at a disadvantage. Those who get drunk together come to trust one another, for all the good reasons documented in Chapters Three and Four. This is why a person who fails to join in is often viewed with mistrust. One ethnographer who did his "field-work" in various pubs in Ireland, studying Irish drinking culture, notes that social acceptance was predicated upon drinking, and drinking hard:

> At one point in my doctoral research I thought it prudent to give up "the drink," and when in pub situations asked for a "soda water and lime." My field notes improved dramatically, but only in line with some deteriorating social relationships. A few key informants, who had become accustomed to shar-ing information with me in pubs, simply wondered why I had gone off the drink, and were suspicious of my motives. I like to tell myself that in the interests of science I was forced to go back on the beer, after a three-month hiatus, after which the relationships that had been in jeopardy were quickly restored.[63]

It is understandable that, in the midst of a group of people mentally disarming themselves, putting their PFCs down on the table, the one person sipping soda water and lime might be frozen out. In the *Game of Thrones* example mentioned earlier, Lord Bolton refrains from drinking because he is keeping his PFC in prime working order so that he can direct the ambush and brutal slaughter of the drunken guests. But what if he just wanted to remain sober in order to take clear field notes and finish his dissertation? What if he were a recovering alcoholic, a Muslim or Mormon, a designated driver, or a single parent who needed to get up early and clearheaded to see the kids off to school?

A helpful piece about the role of alcohol in the tech industry by Kara Sowles, a community manager at a tech company in Portland, Oregon, lays out the inclusivity concerns that arise when work culture is permeated by alcohol, and is worth quoting at length.

In the tech industry, alcohol is currency. It's used to grow event attendance, to bribe participants, to reward employees and community members. Informal interviews are conducted in bars, to see if potential employees are likable in a social setting, or can hold up under heavy drinking with clients. Co-workers gather in pubs to bond and shed the day's frustrations. Good performance is rewarded with shared whiskey, tequila parties, opening up the office taps. We drink to say thank-you, to seal deals, to bid farewell, to make new friends, to rant.

Except...not all of us drink...

There's a myth that people who don't drink are few and far between, because only teetotalers refuse alcohol, and they're a rare bunch (we're not). This myth ignores the multitude of reasons people avoid alcohol. People may not be drinking because they are pregnant—and for many, drinking culture puts them at risk of exposing their pregnancy in situations that could lead to professional discrimination. An increasing number of tech employees are underage, given the industry's fetish for youthfulness, and the rampant use and abuse of interns. They may be on medications that preclude alcohol, and questions as to why they don't drink put them at risk of disclosing their medical history. They might be recovering addicts, trying to avoid alcohol in an industry that places it everywhere without addressing alcoholism or providing adequate support. How about designated drivers, or simply people who are about to drive home? The illusion that "everyone drinks" has no space for the safety of commute.

People might not drink because they're feeling unsafe—understandable in a space where others are increasingly drunk, harassment is common, and alcohol is frequently used to facilitate sexual assault. They might not drink alcohol for religious reasons, and by asking them why they don't drink, you're asking them to reveal their faith. Perhaps they're working early the next morning, or they may be gluten-intolerant and you're serving only beer. They might really be a teetotaler, someone who never drinks alcohol. Or, they may simply not be interested in drinking alcohol that evening.[64]

We have noted the reliance of creative cultures, like the tech industry, on alcohol and other intoxicants. This is useful for boosting both individual and collective innovation and enhancing bonding. However, as Sowles observes, it also comes with costs. Many of the people she has interviewed in the tech industry have reported that the central role of alcohol has made them feel pressured into changing their own drinking patterns in a way that was uncomfortable to them or else face ostracism and social exclusion. Group bonding, by definition, instantly creates outsiders.

Joyful, drunken bonding is a powerful force, but arguably lies behind the continued hold of old boys' clubs in modern institutions. When deals are brokered over booze and cigars in late-night drinking sessions, women and younger men—who shoulder a disproportionate burden of childcaring and household tasks that make attending such sessions difficult—become frozen out of the loop. The same is true when academic expertise is exchanged and collaborations forged over late-night drinks at the conference hotel bar, which women may very well wish to avoid like the plague. Toxic cultural attitudes are more responsible for drunken sexual harassment than the alcohol itself, but this is irrelevant to its potential victims, who simply wish to remain out of harm's way. This means that, whenever alcohol consumption is integrated into a professional environment, those who cannot drink, who choose not to drink, or who feel unsafe around drunk colleagues or superiors become marginalized. This is manifestly unfair, and helps to perpetuate existing hierarchies. Both societies as a whole and individual organizations need to balance the tension between bonding and inclusion, or loyalty and fairness, in ways that will impact their views on alcohol.

The elimination of booze-soaked, informal venues for people to mix and gather does represent a loss of community and channel for honest communication. It also, arguably, leads to healthier livers, lower obesity, and a climate more welcoming to women, non-drinkers, and underrepresented minorities. There is no simple resolution of this tension. The safest strategy is simple prohibition, which is more or less the approach currently adopted, at least in theory, by most companies and other organizations. But this also comes with a cost, as you forgo the benefits of creativity and group cohesion. Getting this balance right requires seeing both sides of the equation clearly.

SOLACE OR WEDGE? REINFORCING BAD RELATIONSHIPS

The anthropologist Dwight Heath's early fieldwork on alcohol and culture focused on the Camba people, an isolated group living near the headwaters of the Amazon in eastern Bolivia with whom Heath interacted over the course of several decades.

Although his experiences with the Camba inspired Heath's views of the importance of alcohol for human sociality, they also provide a cautionary tale in how alcoholic bonding might impede deeper or more healthy social connections. When he first encountered them in the 1950s, the Camba lived unusually isolated lives, eking out economically marginal existences more or less on their own, with family units dwelling in isolated huts. They would come together only on weekends and holidays for almost bizarrely intense drinking sessions, sitting around quietly in a circle taking shots of a viciously potent (89 percent ABV) and rather caustic alcohol produced as a by-product of the local sugar industry, and normally sold as stove fuel. They would drink this vile substance until they passed out, and if they awoke while the drinking was still going on, they'd simply rejoin the ceremony. Heath at the time interpreted this intense level of drinking as a particularly dramatic attempt to forge some sense of group cohesion in an otherwise non-cohesive, atomistic society.

In the 1960s and '70s, the Camba land was connected to more densely populated areas by paved highways and railroads; around the same time, land reforms began breaking up the large sugar-cane estates that previously dominated the region. Inspired by this revolutionary liberation movement, the Camba began to form local peasant unions, or *sindicatos,* cooperating with each other on a much more frequent and intimate basis. Heath found that this new sense of social solidarity, and commitment to a group cause, led to a marked decrease in binge drinking. Later developments, however, served to push the Camba back to their old ways. The increased connectivity to the outside world brought new migrants, who began to dominate economic activity, and new agricultural and industrial practices, like cattle herding and cocaine production by drug traffickers, that destroyed the tropical rain forest ecology and poisoned the local waterways. A military coup resulted in peasant leaders being killed or exiled, and in the abolition of the *sindicatos.* "Lacking the sense of social interrelatedness that they had enjoyed before in the *sindicatos,* then beset by bewildering and damaging changes in all aspects of their own daily lives," Heath observed, the remaining Camba "resumed their earlier patterns of episodic heavy drinking."[65] While the return to binge drinking might have alleviated the psychological pain and anomie to which the Camba had been returned, it arguably also served as a barrier to regaining the more healthy and productive sense of community they momentarily enjoyed during the liberation movement.

Alcohol may play a similarly double-edged role when it comes to personal relationships in industrialized societies. Survey data suggests that married couples who drink together, and in similar amounts, report higher levels of marital satisfaction and have lower rates of divorce.[66] Studies have also shown that drinking together, as opposed to drinking apart, has positive effects on couples' interactions the following day.[67] One way that we would expect modest amounts of alcohol to be helpful to couples would be in resolving conflicts or tensions, with the combination of enhanced honesty, focus on the moment, and elevated mood making it easier to raise and process difficult emotions or deep concerns.

A potential worry, however, is that couples might be using alcohol as a crutch rather than an instrumental aid. This concern is reinforced by a study by Catherine Fairbairn and Maria Testa, which found that romantic couples brought to around .08 percent BAC had better experiences, and treated each other with more generosity and empathy, during a conflict-resolution task. This was true, however, only of couples who, when sober, reported low relationship quality—couples reporting high relationship quality had similarly positive interactions regardless of whether they were sober or drunk. This result, combined with a review of previous literature on the topic, led the authors to conclude that "dissatisfied couples might drink more because they experience more reinforcement from alcohol or, put simply, because they get more out of drinking."[68] This dynamic could put dissatisfied couples at risk of alcohol dependence. The authors note that, when treating couples with acknowledged drinking problems, successful interventions are aimed at increasing both relationship quality and intimacy, which seems effective in reducing a dependence upon alcohol.[69] The temporary chemical connection provided by drink may be both desensitizing or numbing marital couples in unsatisfying relationships and preventing them from doing the work required to forge deeper, more authentic ties.

DRUNK ON HEAVEN: GETTING BEYOND ALCOHOL?

Given these serious concerns about the individual and social uses of alcohol, it is worth considering the ways in which we might get beyond drinking entirely, allowing its functions to be served by other practices. We should recall that the splash of intoxication that makes human life possible, that has allowed our species to solve the problems involved in being the creative, cultural, and communal ape, can be non-chemical in nature—or, at least, not come in the form of an ingested chemical substance. In Chapter Two, we noted that many of the effects of alcohol and other intoxicants, including enhanced mood, loss of sense of self, and reduced cognitive control, can be produced in ways that do not involve the consumption of drugs.

In the New Testament, bystanders are amazed by Christians who have been possessed by the Holy Spirit and have begun speaking in tongues, declaring that they must be drunk. The apostle Peter sets them straight. "These people are not drunk, as you suppose. It's only nine in the morning!"[70] Indeed, the apostle Paul at one point chides the Ephesians, apparently a hard-drinking lot, not to "get drunk with wine, but to be filled with the spirit."[71] A similar story comes to us from an early Chinese Daoist text, the *Zhuangzi*, named after its supposed author. "When a drunken person falls out of a cart," Zhuangzi observes, "although the cart may be going very fast, he won't be killed."

> His bones and tendons are the same as other people, and yet he is not injured as they would be. He was not aware that he was riding, and is equally unaware that he had fallen out. Life and death, alarm and terror cannot enter his breast, which is why he can bump into things without fear. This is because his spirit is intact.[72]

Zhuangzi's goal as a religious teacher is to help people to escape from the control of the conscious mind—if he had had the benefit of modern cognitive neuroscience, he would have identified the enemy more specifically as the PFC. In his view, weakening the hold of the mind allows one to relax into a state of "effortless action," where one can respond to the physical and social worlds spontaneously and authentically, with one's spirit "intact."[73] Zhuangzi sees in an extremely drunken person being carted home from a party at least some version of the desired wholeness of being. The drunk, having lost a sense of self and in the grip of extreme cognitive myopia, is free from self-monitoring, doesn't stiffen up in anticipation of contact with the ground, and therefore emerges unharmed from an accident that would kill a sober person. Nonetheless, the overall thrust of the text makes it obvious that ethanol intoxication is merely a metaphor for a more profound and enduring spiritual state. Zhuangzi wants us to get drunk on Heaven, not on booze. "If a person can keep himself intact like this by means of wine," the story concludes, "how much more so can he stay intact by means of Heaven! The sage hides in Heaven, and therefore nothing can harm him."

We have noted the widespread use of chemical intoxicants by religious traditions around the world and throughout history. It is also worth returning at this point to a discussion of the non-pharmacological methods they have developed for achieving ecstatic states of mind. It is clear that completely "dry" rituals involving dance, especially extended, vigorous dancing, ideally combined with hypnotic music and sensory and/or sleep deprivation, can provide many of the psychological and social benefits of drug-fueled ecstatic group rituals. These practices, of course, should not be seen as non-chemical, in the sense that they are as much a part of the physical chain of causality as a glass of ethanol-containing wine or tab of LSD. As our defender of the "chemicals all the way down" view, Aldous Huxley, puts it:

The chanting of the *curandero,* the medicine man, the shaman; the endless psalm singing and sutra intoning of Christian and Buddhist monks; the shouting and howling, hour after hour, of revivalists—under all the diversities of theological belief and aesthetic convention, the psychochemico-physiological intention remains constant. To increase the concentration of CO_2 in the lungs and blood and so to lower the efficiency of the cerebral reducing valve, until it will admit biologically useless material from Mind-at-Large—this, though the shouters, singers and mutterers did not know it, has been at all times the real purpose and point of magic spells, of mantrams, litanies, psalms and sutras.[74]

The "cerebral reducing valve" of which Huxley speaks is, of course, the PFC, the center of cognitive control and rational focus. His argument is that, despite the diversity of theological views that inform them, the goal of all of these religious practices is physiologically identical: to reduce the activity of the PFC and boost endorphins and other "feel good" hormones, allowing the narrow individual self to be open to the "Mind-at-Large."

If Huxley is right, we should find that non-pharmacological religious practices have the same effects on the body-brain system as alcohol or other intoxicants. This is, indeed, what we see in the few relevant studies available. Perhaps the most interesting of these is a neuroimaging study of the phenomenon of speaking in tongues, or "glossolalia."[75] The subjects were Pentecostal women who had reported experiencing regular, daily episodes of speaking in tongues for several years. In the lab, they were scanned while either speaking in tongues or singing relatively mellow gospel songs, accompanied by music and gentle movement. Compared to the singing condition, the subjects showed "decreased activity in the prefrontal cortices during the glossolalia state." In other words, these Pentecostals seemed to be able to use prayer-induced glossolalia to knock out their PFC as effectively as having downed a few glasses of Chardonnay. We can see here a direct line connecting these women to the early Christians defended by Peter— they were drunk on the spirit, not wine.

Another interesting ritual study demonstrated that "shamanic-type experiences (e.g., dissociation from the body, tunnel experiences)" could be evoked in subjects simply by means of monotonous drumming, especially in individuals who were moderately or highly hypnotizable.[76] In the 1970s, the psychiatrist and spiritualist guru Stanislav Grof developed a technique dubbed "holotropic breathwork," whereby intense hyperventilation is used to starve the brain of oxygen and induce LSD-like experiences.[77] In a review of non-chemically induced "hypnagogic states," or episodes of dreamlike disassociation from waking reality, the psychologist Dieter Vaitl and colleagues[78] list a variety of techniques by which such states can be induced, including extreme temperatures, starvation and fasting, sexual activity and orgasm, breathing exercises, sensory deprivation or overload, rhythm-induced trance (drumming and dancing), relaxation and meditation, hypnosis and biofeedback.

All of this will sound very familiar to any student of religious history and comparative religion. In the Sufi tradition, for instance, the so-called "whirling dervishes" employ intensely exhausting dance and hypnotic music to produce ecstatic religious states.[79] Revivalist American Christianity in the late-eighteenth and early-nineteenth centuries was characterized by alcohol-free, large-scale events that featured ecstatic practices and collective effervescence. For instance, the Cane Ridge Revival, held in Kentucky in August of 1801 and dubbed "the first Woodstock" by cultural commentator Harold Bloom, was an enormous camp meeting that brought together between 10,000 and 20,000 people over the course of a week. It featured multiple stages and preachers, dancing, and song, plus religious "exercises" such as "falling," "jerks," "barking," and "running"—all without the aid of ingested chemical intoxicants.[80] The culture of the !Kung hunter-gatherers in southern Africa includes intensely bonding fireside chats and healing trances brought about through ritual singing and sleep deprivation, again in the absence of alcohol or other chemical intoxicants.[81]

So, religious traditions across the world and throughout history have been able to tap into practices that can produce many of the individual and social benefits of chemical-substance intoxication without the toxins. Given the various costs and harms involved with chemical substances like alcohol, one might wonder why they haven't been completely displaced by non-toxic substitutes. However, as we observed in Chapter Two, this is probably because these alternative practices are physically exhausting, often difficult and/or painful, and awfully time-consuming. Given a social goal—say, feeling a bit euphoric, opening up to my friends, and emerging more closely bonded to them—and a menu of techniques for getting me there, the rational choice might very well be to pick a two- to three-hour drinking session over an all-day ritual involving intense physical exertion, physical pain, or both. Insights provided by a five-hour mushroom trip might be just as valuable as those dragged out of my unconscious by a three-day silent retreat. Moreover, staying up all night, sticking sharp stakes through my cheeks, or dancing or meditating all day instead of harvesting the crops impose their own distinct costs on both individuals and cultures.

Humans have invented a wide variety of methods for downregulating the PFC, upregulating mood, and helping people to be more creative and open, and they all have their own specific costs and benefits. Why particular cultures have settled upon techniques that do not involve the ingestion of chemical intoxicants may be purely a chance result of random cultural variation, or could be driven by particular, local trade-offs involving the relative costs and benefits of various methods.

That said, if the costs of any particular technique increase, or if it's discovered that its results can be achieved in less costly ways, we would expect that technique to be gradually displaced. For instance, the historian of Chinese religion Gil Raz observes that in one particular sect of Chinese Daoism, the practice of using psychedelic herbs to achieve states of ecstatic insight and unity with the sacred was gradually replaced by guided meditation and complex breathing techniques, which were able to produce similar spiritual results without potentially dangerous side effects.[82] Similarly, if it's true that distillation and isolation have recently ramped up the potential dangers of alcohol consumption, this could provide a novel competitive advantage to cultural groups that eschew it. It is entirely possible that the relative success, in recent centuries, of Islam, Mormonism, and alcohol-rejecting forms of Christianity has been driven, at least in part, by this dynamic.

TAMING DIONYSUS

While we await the worldwide takeover by teetotaler forms of religiosity, and the wholesale replacement of bars and pubs with holotropic breathing stations, alcohol and related drugs will continue to be our method of choice for dialing down the prefrontal cortex and enhancing our creativity, cultural openness, and communal feelings. Dionysus is undeniably dangerous. He can turn you into an animal, or give you a gift, like the golden touch of Midas, that turns out to be a curse. Considering the dangers and costs involved in allowing chemical intoxication to play a role in our lives, prudence would dictate that we think about how to mitigate the risks. Entire books can and have been written about precisely this topic.[83] Here, I would like to conclude this chapter with a few takeaways that emerge naturally from our discussion.

Sober Bars: Tapping into the Placebo Effect

Ruby Warrington, a journalist who came to view with alarm the toll that regular work-related drinking was taking on her health and mind, has organized a movement for what she calls the "sober curious," which hosts events and retreats. She is part of a broader movement of what has been described as:

> a new generation of kinda-sorta temporary temperance crusaders, whose attitudes toward the hooch is somewhere between Carrie Nation's and Carrie Bradshaw's. To them, sobriety is something less (and more) than a practice relevant only to clinically determined alcohol abusers. Now it can also just be something cool and healthful to try, like going vegan, or taking an Iyengar yoga class.[84]

This movement has given rise to "sober bars," like Getaway in Brooklyn, where people can socialize in a bar-like setting while drinking delicious and interesting virgin cocktails. One way of understanding sober bars is as a method for tapping into the intoxication buzz without the toxic part.

We've mentioned before the expectancy effect that surrounds alcohol consumption. If you drink something *expecting* that it is going to make you drunk, it often makes you a little drunk, even if it's only flavored water. This is related to the well-known placebo effect in medicine; patients who are given a sugar pill and told it is a potent medicine will often see significant health improvements. More relevant to the phenomenon of sober bars, merely *thinking* about alcohol, by being primed with alcohol-related keywords or shown an alcohol-related ad, can make you feel and behave a bit drunk.[85] So patrons of a sober bar, despite knowing they are being served virgin cocktails, are nonetheless exposed to unconscious alcohol-related cues. Sitting in a bar-like setting, with the lights dimmed and music playing, and being served drinks that look and taste very much like alcoholic cocktails can therefore provide many, if not most, of the social benefits of drunkenness without the costs. One team of researchers took advantage of expectancy effects to demonstrate to at-risk college drinkers that they could have just as much fun in an environment where they thought they were getting drunk as when they were served actual alcoholic drinks, which in turn helped to train them to socialize without the need for alcohol.[86]

The power of this effect has caused some commentators to conclude that the psychological and behavioral effects of alcohol are *all* produced by cultural expectations. This is especially the case in fields like cultural anthropology, where the dominant theoretical models see human experience as completely socially constructed from the ground up.[87] As we have noted above, however, it is clear that cultural expectations about alcohol are very much dependent upon, and driven by, the actual pharmacological effects of ethanol. It is not simply a coincidence that similar cultural expectations, throughout history and across the world, are attached to alcohol. Drunkenness is conceptualized in ancient China and ancient Egypt and ancient Greece in quite similar ways because it is the product of the same chemical hitting the same sorts of brain-body systems. It has also become clear, since the advent of balanced placebo designs, that expectancy effects are not as powerful as some early research seemed to suggest. The ability to tease apart the results of thinking you are drinking when you are not and thinking that you are *not* drinking when you are has demonstrated that many psychological and behavioral outcomes are, indeed, being driven by the pharmacological effects of alcohol.[88]

This suggests limits to the ability of sober bars to genuinely enhance sociality beyond simply physically bringing people together in an environment that encourages relaxation and conversation. Sober bars derive their power from, and are therefore dependent upon the existence of, "real" bars that serve alcoholic cocktails that actually do ramp up your BAC. In a world where nothing but near-beer and virgin cocktails existed, the cultural concept of alcohol would gradually lose its power.

Mindful Drinking

In her helpful and funny book, *Mindful Drinking*, Rosamund Dean makes the important observation that much of our alcohol consumption is "mindless," in the sense that we habitually pour ourselves a glass of wine at the end of the day, or accept another proffered drink at a reception, without considering if we actually really want it. Simply pausing and making a conscious decision about whether or not we do, in fact, want that drink can go a long way toward moderating consumption. She offers a useful set of principles for those who would assume better control of their drinking habits, what she calls "The Plan," based on conscious monitoring of one's alcohol intake and drinking only when it can be savored and enjoyed.

In addition to adopting a general attitude of moderation and mindfulness, there are some simple tricks we can employ to keep alcohol consumption within modest limits. One of my favorites comes from ancient Greece, where, as the classicist James Davidson notes, wine cups were deliberately shallow, so using them without spilling required significant motor control.[89] This, in turn, served to indirectly limit wine consumption once BAC rose beyond a certain point. Also, in the same way that serving food on smaller plates reduces overeating, reducing the size of beer and wineglasses helps people to regulate their consumption. So does mixing in alternating rounds of soda water or other non-alcoholic drinks when out drinking or drinking at home. With regard to drinking at the workplace or in professional contexts, there are obvious sensible limits that can be imposed: no completely open bars, employ drink tickets and drink limits. Most organizations already have such guidelines in place, although in some cases they were only implemented after work-related drinking had gotten completely out of hand. You ought to know your company has a problem long before you're finding cigarettes, beer cups, and used condoms in the office stairwells, which is what it took for one tech company to finally modify its workplace guidelines.[90]

Beware Liquor and Don't Drink Alone

We are apes built to drink, but not 100-proof vodka. We are also not well equipped to control our drinking without social help. Distilled spirits are so much more powerful and dangerous than beer and wine that they should always be treated as a separate drug, and regulated accordingly. People under twenty-five should avoid them. People over twenty-five probably should too, but at least their PFCs are fully developed, so they might have a leg up on making reasoned decisions about what to do with their body-brain systems. In any case, hard liquor bans on college campuses, which have been adopted by many institutions, seem like a pretty smart policy. Taxing liquor at a much higher rate than beer and wine, and limiting its sale, is also quite sensible.

The role of toast leader or symposiarch in ancient Greece was a critical and honored one. This figure was required to evaluate participants' state of inebriation, pace the drinking, and have those who had gone over the line sent home. In the modern world, something like this role can and should be played by bartenders and cocktail servers, especially when it comes to people drinking alone in pubs or bars. There are some concrete structural reforms that can be made to allow these modern symposiarchs to do their jobs more effectively. In the United States, for instance, alcohol servers should get paid real base salaries, as they are in Europe and Asia. This would help eliminate the tension between their role as social monitors—which requires that they regulate the pace of drinking, cut off excessively inebriated customers, and confiscate car keys—and their current dependence on tips, which penalizes them for not indulging patrons' every whim. I spent my college and grad school years working in restaurants, bars, and night-clubs in San Francisco. As a server I was paid only an "alternative minimum wage," below $2.00 an hour, by the restaurants and bars I worked in, and was therefore completely dependent on tips to survive. This created a perverse incentive system. I can recall many evenings when my feeling that a certain table or patron had already had enough to drink was trumped by my fear of getting stiffed if I cut them off. Alcohol is also a very easy way to run up a tab and thereby increase the gratuity. I would predict that a modest investment in salary raises for bartenders and servers would immediately slash drunk driving deaths, public fighting, and a host of other ills.

A well-trained and thoughtful bartender is no help, of course, to someone drinking at home. Many homes are stocked with such a generous supply of wine, beer, and spirits that one can, like a rat in a stress study, suckle on alcohol more or less without limit, except perhaps for the guilt-inducing speedbumps of having to uncork a second bottle. Hypnotherapist Georgia Foster runs a practice in the U.K. aimed at helping clients, mostly women in their thirties and forties, manage their drinking habits. She notes how dangerous drinking at home, throwing a "solo party" for one-self, can be for single women or stay-at-home moms: "When you're in your own environment, there's nobody watching and you don't have to drive anywhere, that glass of wine tends to go to two, then three, then the bottle's gone and you open a second bottle. That can be a very slippery slope."[91]

Again, we are simply not well adapted, evolutionarily, to be able to consume alcohol safely outside of the traditional context of ritual and social controls. One possible solution would be to simply decide to drink only in public, with other people, under the watchful eye of your local pub's barkeep. Or, if consuming at home with your family over meals, to limit drinking to the dinner table, as is the case in Southern drinking cultures like Italy or Spain. For those who live alone, virtual social drinking is now possible through apps that allow anyone with a smartphone to tap into social feedback while drinking. Some early studies utilizing apps that connect otherwise solitary drinkers to a social network show promising results in terms of helping them to moderate their consumption.[92]

Normalizing Alcohol: Spreading the Southern Model

We have noted that, speaking in terms of European geography, so-called Southern drinking cultures provide some protection against the novel scourges of distillation and isolation. The pathological aspects of Northern drinking culture, in contrast, have probably reached their peak in the United States, where they are reinforced by the particularly American strand of Puritanical bipolarism when it comes to pleasure. Drinking in American culture is even more compartmentalized than in Northern Europe. Alcohol is more rarely consumed with meals and is much more likely to be demonized. It is no accident that the United States is the only non-Muslim industrialized nation to attempt to prohibit alcohol altogether. As the anthropologist Janet Chrzan notes, even today the U.S. has the highest rate of self-reported abstinence, around 33 percent, in the non-Muslim world. This is several times greater than classic Northern drinking cultures such as Sweden (9 percent) or Norway (11 percent). Americans' weirdly conflicted attitudes toward booze is most pronounced in socially and religiously conservative "red" states. Chrzan relates a story from a period she spent in a rural region of South Carolina. She would occasionally run into someone she recognized at the local liquor store, but found she was completely ignored when she tried to greet them. After this happened several times, she consulted a local friend, who laughed and said, "You are sure not from here! Don't you know you *never* say hello to Baptists in the liquor store?" If you're a Baptist in South Carolina, you can buy and drink alcohol, but only if you keep it secret, and drink it in private.

As Chrzan argues, this fraught relationship to alcohol makes America an "all or nothing drinking nation," swinging wildly between abstinence and extreme indulgence, even to the point of winking at or overlooking violent or sexual excesses that would otherwise be condemned.[93] At American universities, this extreme Northern attitude, combined with the freedoms of college life and an advertising industry that celebrates drinking as something cool or heroic, produces a "cultural trifecta almost perfectly designed to encourage abuse among youth."[94]

Northern attitudes toward alcohol have a tendency to lead to problem drinking. As Dwight Heath notes, in societies where alcohol is not well integrated into daily social life, it can take on a "mystique of holding the implicit promise of imparting power, sexiness, social skills, or other special qualities," thereby motivating people, especially young people, to "drink too much, too fast, or for inappropriate and unrealistic reasons."[95] To be fair, drinking to get drunk has always been part of the appeal of intoxication: The bacchanalia depicted on the cover of this book is not a carefully regulated dinner party, and the original Greek followers of Dionysus were not known for pausing for soda water in between rounds. But in healthy cultures, drinking to get drunk is rare, and typically only indulged in as part of specific, sanctioned rituals, such as carnival. Outside of these contexts, drinking is moderate, focused mostly on wine and beer, and public drunkenness seriously frowned upon.

One way to instill "Southern European" drinking attitudes no matter where you live is to introduce young adults to moderate drinking practices, in the context of meals, at home. My own daughter, now fourteen, is allowed to smell and taste small samples of whatever wine I am drinking and has already developed a fairly sophisticated palate, able to reliably pick out apricot or lemon notes in a Chardonnay. The point is not to turn her into an insufferably pretentious snob, but to introduce her to the idea that wine can be a source of aesthetic pleasure. This is much preferable to seeing it as some forbidden substance used solely by adults to get drunk. She has also had the health risks clearly explained to her, and understands that she is too young to drink wine in any quantity. This is not because it is an exotic taboo elixir, but because her PFC is nowhere near as big as it needs to be, and throwing hand grenades at it now would be pretty stupid from a developmental perspective. It doesn't hurt that half of her family is Italian, so she has spent a fair chunk of her childhood in Italy, being exposed to moderate Southern drinking habits. Hopefully this will set her up to develop into a responsible adult who can enjoy alcohol in its proper place, for the proper reasons, and avoid whatever binge-drinking subcultures she will encounter at college.

In this context, it seems sensible to have a lower legal drinking age for wine and beer, perhaps with special conditions for young adults drinking with parents. In the same way that a young person with a learner's permit can drive only in the daytime, with a responsible adult in the passenger seat, one could imagine an older teen being allowed to legally consume a small amount of wine in a restaurant in the context of a family dinner. On the other hand, as mentioned above, the legal drinking age for hard spirits should be substantially higher than it currently is in most jurisdictions.

Try to Level the Playing Field for Drinkers and Non-Drinkers

This is perhaps the most challenging recommendation to implement. If alcohol continues to play a crucial, functional role in social and personal contexts, it is hard to know how to accommodate non-drinkers. For instance, if a tequila-shot-soaked final bender at a dive bar does serve an important and irreplaceable function in bonding together a team of Navy Seals, a devout Mormon may be left out in the cold. If tossing ideas around at a conference hotel bar fosters new collaborations and innovations, it seems counterproductive to close the bars down, even though this means that non-drinkers or those leery of the environment are thereby disadvantaged.

Kara Sowles, the tech company community manager quoted above, concludes her piece on how drinking culture can undermine inclusivity with some concrete suggestions about how to integrate alcohol into professional gatherings without thoroughly marginalizing non-drinkers. Her five pointers are:

1. Provide an equal number and quality of alcoholic and non-alcoholic drink options.
2. Display alcoholic and non-alcoholic drinks together at the event.
3. Advertise alcoholic and non-alcoholic drinks equally before the event.
4. If listed cocktails are being served, list an equal number of non-alcoholic mocktails.
5. Have water freely available, in clear sight, and easy to obtain.[96]

The last recommendation, in particular, is incredibly important and useful—people without easy, obvious access to water are almost guaranteed to overdrink in a typical reception or social mixer environment. Following simple steps such as these can go a long way toward allowing an organization to capture the benefits of alcohol-enhanced socializing while minimizing the costs to non-drinkers.

Doing so may become increasingly important. Non-drinkers like Sowles are becoming more and more common among young people in industrialized societies. A recent survey of U.K. millennials, for instance, showed that abstinence is becoming more "mainstream," with the percentage of teetotalers among sixteen- to twenty-four-year-olds rising from 18 percent in 2005 to 29 percent in 2015. Binge drinking also seems to have become less socially acceptable, and there is less stigma attached to being a non-drinker in social settings where others are drinking.[97] More generally, there appears to be a worldwide trend among millennials and Generation Z toward either total abstinence or carving out alcohol-free periods of the year, such as Dry January.[98] Leveling the playing field is therefore likely to become an increasingly pressing concern.

LIVING WITH DIONYSUS

We always need to keep in mind the danger of the rampaging maenads, wildly drunk and tearing apart with their teeth any person unfortunate enough to cross their path. In order for the calculus in favor of intoxicant use to remain viable, the potential costs need to be mitigated, and the dangers guarded against. Doing so is particularly challenging in our modern world, where we face the novel perils of distillation and isolation.

It may seem odd to have the final chapter of a book intended to celebrate the functional importance of alcohol focus instead on how and why booze can be bad for us. Acknowledging the downsides and risks of chemical intoxication, however, is crucial to any comprehensive defense of ecstatic chemical joy. And chemical joy *needs* to be defended. As helpful as they are in terms of practical suggestions for moderating drinking or adopting abstinence, most of the "new sobriety" self-help books cited above portray alcohol as an unmitigated vice, a mind-hijacker imposed upon us by greedy corporations abetted by clever marketing companies. Too much of the contemporary literature on alcoholism or problem drinking shares this view. It is a mildly ascetic one, in which alcohol might be allowed a small place in an otherwise healthy, mindful lifestyle, but only rather grudgingly, as a sop to our inherent weakness for pleasure or as an occasional, guilty reward for having attended a certain quota of sunrise yoga classes.

This is historically myopic and scientifically unsound. At the end of the day, despite the potential chaos he brings in his wake, we should welcome Dionysus into our lives. We should do so partly in recognition of the challenges we face as a species and the functional benefits he can continue to provide us only partially civilized apes living in artificial modern hives. We should also accept the fact that pleasure is good, for pleasure's sake, and requires no further justification. We need to debunk the view that alcohol and similar intoxicants are sinister inventions of capitalist modernity, sold to us by nefarious advertisers, leading to nothing but hangovers, lost income, and bulging waistlines. Drinking *can* make us fat, harm our livers, give us cancer, cost us money, and turn us into useless idiots in the mornings. It can even be lethal, to both us and those around us. It has, nonetheless, always been deeply intertwined with human sociality, and for very good evolutionary reasons. Moreover, its important functions are very difficult, if not impossible, to replace with other substances or practices. Let us therefore embrace Dionysus with appropriate caution, but also with the reverence that he is due.

CONCLUSION

Jesus has a problem. He's at a wedding in Cana, with his disciples and his mother, and the wine has run out. People are panicking. His mother gives him a nudge with her elbow and a significant look: *Hey, Son of God, do something about this.* Jesus is reluctant, he wasn't planning on revealing his true nature quite yet, but this is an emergency. *The wine has run out.* So he gets the servants to fill some huge stone containers with water. Then Jesus transforms the water into a wine so exquisite that, when it is delivered to the steward, it is proudly announced that the decision has been made to violate the first rule of party planning, which is serve the good stuff first. The disciples are astounded, having witnessed Jesus' first miracle. Everyone else is just happy to have more wine and the wedding merrily proceeds.[1]

Of course, Jesus goes on to perform several more impressive miracles, including walking on water and raising Lazarus from the dead. But it is worth noting that the water into wine feat was his first. Alcohol is so fundamentally intertwined with human sociality that it pressured the Son of God into his first miracle. And let's not even get into the Eucharist and blood of Christ thing. After our necessary and important exploration of Dionysus' dark side in Chapter Five, it is time to return here to the main thrust of this book, the joy and power of intoxication.

As we've seen, alcohol takes on a sacred quality in most cultures. Texts from medieval China explain how the water to make "divinely fermented wine" can only be gathered before sunrise on a particular day of the month, by a ritually purified boy in a very specific manner, and then cannot be touched by another human hand.[2] As a sacred substance, alcohol is also often seen as possessing magical powers, or conveying such powers on those who imbibe it. In the earliest written text from Japan (the *Kojiki*), the emperor rhapsodizes over the "august heavy liquor" brewed by a visiting Korean noble who, legend says, introduced a form of sake to Japan. "And oh, how drunk I am! / On the evil-chasing, / Laugh-giving liquor—/ And, oh, how drunk I am!"[3] Upon wandering, happily inebriated, out of the palace, the emperor strikes with his staff a stone that is blocking his path; the stone promptly jumps up and runs away. In Mexico, *pulque,* considered a sacred beverage in precolonial times, was in the Christianized era dubbed the "milk of our Mother" (Mary), sacrificed to souls on the Day of the Dead, and poured upon skulls buried in the four corners of fields to protect from robbers.[4] Throughout Africa, the numinous power of beer is seen as the essential component of religious ceremonies and sacrifices to the ancestors. The Kofyar of northern Nigeria believe that "[people's] way to god is with beer in hand."[5] As one Tanzanian puts it, "If there is no beer, it's not a ritual."[6] Because of its special status, cultures often define themselves in terms of specific alcoholic beverages—think of the French and wine, Bavarians and beer, and Russians and vodka. As the anthropologist Thomas Wilson notes, "In many societies, perhaps the majority, drinking alcohol is a key practice in the expression of identity, an element in the construction and dissemination of national and other cultures."[7]

These sacred or culture-defining beverages differ wildly in production method, color, taste, and consistency. What they have in common is possessing ethanol as an active ingredient. Why is this particular neurotoxin accorded such reverence? Because alcohol, pre-eminently among the chemical intoxicants employed by human beings, is a flexible, broad-spectrum, and powerful technology for helping us to inhabit our odd, extreme ecological niche. We wouldn't have civilization as we know it without intoxication in some form, and alcohol has been, far and away, the most common solution that cultures have hit upon to fill this need. In addition to its social functions, intoxication is also a badly needed salve for the only animal on the planet afflicted with self-awareness. "We are chimps with brains the size of planets," laments Tony, Ricky Gervais's character in the lovely show, "Afterlife." "No wonder we get drunk."[8]

Explicitly acknowledging and documenting the functional usefulness, individual solace, and deep pleasure provided by alcohol and other intoxicants is a much-needed corrective to today's popular wisdom on the topic. Intoxicants are not merely brain hijackers or vices to be eliminated or grudgingly tolerated. They are essential tools in our battle against the limiting aspects of the PFC, that seat of Apollonian control, as well as the constraints of our primate nature. We cannot properly grasp the dynamics of human social life unless we understand the role that intoxicants have played in making civilization possible. As that great champion of Dionysus, Friedrich Nietzsche, declared in one of his characteristically cryptic aphorisms, "Who will ever relate the whole history of narcotica? It is almost the history of 'culture,' of our so-called higher culture."[9]

This book was mostly written in the midst of the Covid-19 crisis, which has provided dramatic confirmation of the ineliminable role of alcohol in our lives. One of the big debates early in the pandemic, when governments were imposing lockdowns, was what counted as an excepted "essential service"? There was wild and bizarre variety in the answers to this question across the United States. Some states declared golf courses exempt, others gun shops. One thing that every jurisdiction recognized, however, and that never seemed up for debate, was that liquor stores are essential. (The one state that attempted to close liquor stores, Pennsylvania, quickly changed course in response to public outrage.)[10] In Canada and the U.S. states where it is legal, the essential services exemption also extended to recreational cannabis dispensaries. It is also worth noting that the few countries that used Covid-19 as an excuse to attempt prohibition, like Sri Lanka, ended up spawning enormous underground networks of home brewers, cooking up barely palatable—but definitely intoxicating—concoctions out of everything from beets to pineapples.[11] People want to drink, and even a global pandemic will not stop them from doing so.

Understanding why is profoundly important. This question cannot be coherently asked or answered without understanding the function that alcohol has played in human civilizations. As we have seen, besides its immediate hedonic value, the cognitive and behavioral effects of alcohol intoxication represent, from a cultural evolutionary perspective, a robust and elegant response to the challenges of getting a selfish, suspicious, narrowly goal-oriented primate to loosen up and connect with strangers. To have survived this long, and remained so central to human social life, intoxication's individual-level advantages, combined with group-level social benefits, must have—over the course of human history—outweighed its more obvious costs. This is why both genetic and cultural "solutions" to the alcohol "problem" have failed to spread as quickly as one would expect if our taste for intoxication were merely an evolutionary mistake.

What this means in our modern world, massively complex and

changing at an unprecedented rate, is something we can only evaluate properly when we take a broad historical, psychological, and evolutionary perspective. Doing so might lead us to conclude that, when it comes to achieving certain goals, consuming alcohol should be replaced by better, safer methods. Non-alcoholic alternatives may be particularly appealing in an age where we face the relatively novel dangers of distillation and isolation. For instance, if the aim is to enhance group bonding or sense of being part of a team, it might very well be the case that a company outing to the laser tag park or an escape room will provide the same outcomes as an alcohol-soaked holiday party, with none of the downsides. As we gather more data on microdosing psychedelics, we might find that they provide all of the creativity-boosting power of alcohol without the danger of addiction or liver damage.

Other cases are more contentious and complicated, but even here it is useful to provide a scientifically accurate outline of the decision-making landscape. Maybe office parties should be alcohol-free, or held in the mornings with a one-mimosa limit. Perhaps it is good and right that Canadian federal grant funds—even those specifically dedicated to networking—cannot be spent on alcohol. What are the costs and benefits of limiting or eliminating alcohol in this way? To be sure, drinking in moderation seems less controversial than getting truly drunk (above .10 percent BAC), but is excess always bad? The landscape here becomes more tangled and confused. Excess is clearly dangerous, leading to exponentially increasing costs, but is not necessarily categorically maladaptive. Well-timed excess can sometimes be useful in bonding certain types of groups together or helping individuals move past difficult moments in a relationship. It must be the case that, over the course of our evolutionary history, the ability to chemically disarm, to honestly make oneself vulnerable to another, has provided an essential social benefit to outweigh the obvious costs.

At the very least, when it comes to the science it is time to move beyond evolutionary hijack or hangover theories of intoxication and, when it comes to cultural attitudes, to overcome our outdated folk notions and moral discomfort. Debates about the proper role of intoxicants in our lives need to be informed by our best current scientific, anthropological, and historical scholarship, which at the moment is far from the reality. Getting the right perspective will put us in a place to more clearly see what concrete trade-offs we face when we formulate policies and make personal decisions about the role of intoxicants in our lives. Our desire for alcohol is not an evolutionary mistake. There are good reasons for why we get drunk. No informed decision, at either the individual or social level, can be made without a better appreciation of the role that intoxication has played in creating, enhancing, and sustaining human sociality, and indeed civilization itself.

Doing so, however, is particularly challenging in today's climate, which is simultaneously technocratic, ascetic, and moralistic. A recent *Lancet* article on historical and projected alcohol use[12] notes that global adult per capita consumption of alcohol increased from 5.9 liters to 6.5 liters between 1990 and 2017, with lifetime abstinence decreasing from 46 percent to 43 percent. The authors predict that these trends will continue, with abstinence decreasing to 40 percent by 2030. The article's conclusion, not presented as an interpretation but simply as a fact, is that this is clearly a public health disaster in the making, and that we need to mobilize all known policy measures to reduce alcohol exposure and reverse this trend. This attitude only makes sense against a backdrop where anything that doesn't directly increase our individual life-span or lower our cancer risk is categorically bad. This modern, secular version of asceticism, whether grounded in medical advice or the teachings of relentlessly mindful lifestyle gurus, also informs much of the contemporary self-help literature on alcohol use.

There is little room here for broader, long-term considerations of what allows human beings to live and create together in productive civilizations, or what gives life texture and makes it enjoyable and worth living.

Perhaps a deeper problem is that our age is moralistic to a degree not seen since Queen Victoria's day. In part, this is a long-overdue and important corrective to laissez-faire attitudes that left oppressive gender norms and racial prejudices intact, winked at 1950s-style *Mad Men* behavior, and embraced a retrograde "boys will be boys" attitude. At its most stifling, however, the new moralism makes it difficult to discuss clearly and objectively certain topics at the center of human experience. Chemical intoxication is rivaled only by sex as a taboo topic. It is largely overlooked by scholars of human sociality, and its benefits are ignored in public policy decisions. As Stuart Walton complains:

> Intoxication plays, or has played, a part in the lives of virtually everybody who has ever lived, and yet throughout the entire Christian historical era in the West, it has been subject to a growing accretion of religious, legal and moral censure. These days, we are scarcely able to whisper its name for fear of falling foul of the law, of compromising ourselves as being part (however peripherally) of the multiple blight that has befallen our societies in the form of cigarette-smoking, drink-driving, hooliganism, self-inflicted illness or drug-related crime.[13]

We need to rescue alcohol, and chemical intoxication more generally, from both the cheery New Age ascetics and dour neo-Puritans.

My nagging worry is that I have only half-accomplished this, by couching my defense of intoxication in pragmatic, functional terms—talking about costs and payoffs, and framing everything in terms of an evolutionary calculus. My hope is that I have instead managed a holistic defense of alcohol and intoxication, one that also gives pleasure for pleasure's sake its due. It might be helpful, in this regard, to return to Tao Yuanming, a writer from whom I probably first derived, as a young scholar of Chinese, my interest in the topic of intoxication. This is number 14 of his wonderful "Drinking Wine" series:

Old friends appreciate my tastes;
So they arrive carrying a jug of wine to share.
We spread our mats and sit under a pine tree,
And after a few cups are thoroughly drunk.
Just some old guys rambling from topic to topic
Losing track of whose cup should be filled next.
Completely free of self-consciousness,[14]
Why should we value one thing over another?
So relaxed and distant, lost in where we are;
In the midst of wine there is profound meaning![15]

The last two words of the poem are very difficult to render smoothly or accurately into English. *Shenwei* 深味 means literally "deep/profound taste/flavor/meaning," and includes both spiritual and hedonic overtones. Wine is both meaningful and pleasurable.

It seems appropriate to close this book by recounting our earliest myth about Dionysus, in a Homeric hymn likely dating to the seventh century BCE.[16] Appearing in the form of a youthful and well-dressed man, the god is captured by pirates, who assume that he is a wealthy ruler's son who can be ransomed for a handsome price. Only the helmsman is worried about this plan, because he recognizes Dionysus as a god and is appropriately awestruck. Once the group sets out to sea, all sorts of magical hell breaks loose. The ocean turns to wine, the mast transforms into an enormous, grape-laden vine, and, finally, Dionysus takes on the form of a lion as the terrified sailors flee into the water, where they are turned into dolphins. Only the helmsman, to whom Dionysus finally reveals his true identity, is spared. He goes on to live a long and prosperous life, having been personally blessed by the god.

This story is wonderful and revealing. "Few recognize Dionysus as a god, this hymn seems to tell us," the classicist Robin Osborne observes, "and only those who do retain their humanity."[17] It is also a fitting way to end our account. The ancient Greeks despised "water drinkers," whose rejection of wine reflected a cold heart, plodding mind, and even moral turpitude. Today we are, rightly, more aware of the value of abstention, and are not likely to re-establish Dionysus in our official religious pantheon anytime soon. However, it is only by recognizing both the benefits and dangers of intoxication that we can remain human, cautiously tapping into its power to help us to occupy this precarious ecological niche we have made our own. As the hymn to Dionysus concludes, "Hail, child of fair-faced Semele! He who forgets you can no longer create sweet song."[18] Let us retain our humanity, being sure not to forget Dionysus, but rather to see him clearly as both a god and a threat. It is only in this way that we can leave a place for ecstasy in our lives, preserve the ability to "create sweet song," and continue to flourish as humans, the strangest, and most successful, of apes.

ACKNOWLEDGMENTS

Like all humans, I would be as helpless without communal support as a blind cave tetra thrown into a sunlit stream. I have been thinking about this project, and talking about it with others, for so many years that I am absolutely guaranteed to overlook multiple people who shaped my thinking on the topic or guided me toward relevant resources. My apologies in advance.

The members of our core research team at the University of British Columbia, Joe Henrich, Ara Norenzayan, Steve Heine, and Mark Collard, all played important roles—Joe, especially, during our island kayak camping trips and real and virtual pub sessions. Michael Muthukrishna gave me careful, thoughtful feedback on Chapter Two, suggested helpful references throughout, and over the course of the writing sent me a steady stream of relevant popular press and academic publications. Tommy Flint at Harvard provided crucial help in the early stages of research, guiding me through the ADH/ADLH literature, questioning my reasoning about genetic and cultural evolutionary dynamics, alerting me to the wonderful "Bar Talk" economics paper draft, and providing a host of other suggestions, including the relevance of "night talk" among the Baku. Emily Pitek at Yale conducted the HRAF database survey reported in Chapter Three; many thanks for her hard and thoughtful work on this project. Michael Griffin generously walked me through his understanding of the early Homeric hymn to Dionysus with which I conclude the book.

Thanks, too, in no particular order, to Hillary Lenfesty (pentecostals and speaking in tongues); Chris Kavanaugh (Shinto ritual and animal intoxication); Will Gervais (wide-ranging discussions over Kentucky bourbon); Randy Nesse (great dinner conversation about intoxication, helpful catch on an error in the manuscript and reference recommendations); Bob Fuller (for generously sending a copy of his very helpful book on religion and wine); Sam Mehr (music and intoxication); John Shaver (kava); Polly Wiessner (!Kung and non-chemical intoxication); Sarah Pennington (holotropic breathwork); Gil Raz (Daoist practices); Willis Monroe and especially Kate Kelley (leads on Mesopotamian beer); Amanda Cook and Pico Iyer (book recommendations); Jan Szaif; Leanne ten Brinke; and Nate Dominy (helpful articles). Dimitris Xygalatas provided helpful comments on extreme rituals and sent me useful reading material, and Alison Gopnik generously spent some time in Arizona walking me through her work on childhood and creativity. I also received helpful feedback from Nathan Nunn, Lucy Aplin, and the Twitter hive mind that helped me track down that Sarah Blaffer Hrdy citation.

I am also very indebted to Robin Dunbar, both for his groundbreaking work on this topic and for generously sharing relevant preprint articles with me, and to Michael Ing for his chapter in progress on Tao Yuanming and a slew of extremely useful references on attitudes toward drinking and drunkenness in early China. Thanks to Jonathan Schooler for discussions of mind wandering and wide-ranging chats on chemical intoxication and insight, and Azim Shariff for general encouragement and recommendations of literature and structural suggestions.

Ian Williams, a gifted poet and writer and my longtime tennis partner, provided essential between-game conversation, emotional support, and words of wisdom about the craft of writing and life in general. (The serve is looking awesome, by the way.) Michael

Sayette provided some helpful references early in the writing process and then very generously gave me an enormous amount of thoughtful and constructive feedback on the final draft, ranging from high-level theoretical considerations to detailed copyedits. If I'd engaged more with Michael earlier on this would have been a much better book; in any case, I did my best to supplement areas of weakness and bolster arguments where I could.

Deep love and thanks, as always, to Stefania Burk and Sofia, for many things, but most of all for showing me the strength, importance, and happiness of family. Stefania also commented on some early chapters, and Sofia, as usual, unwittingly provided a host of useful real-life examples. Tracking down my sometimes snarky references to her may end up being sufficient motivation to read one of her dad's books for the first time.

Marcos Alberti very generously gave me permission to use some photos from his wonderful "3 Glasses Later" project, Dick Osseman allowed me to use his image from Nevali Çori, Randall Munroe gave permission to reprint his xkcd cartoon, and Kara Sowles allowed me to quote extensively from her essay on leveling the playing field for drinkers and non-drinkers. Thanks also to the German Archaeological Institute (DAI) for making their images of Göbleki Tepe available for reuse.

I'm grateful to my agent at Brockman Inc., Katinka Matson, and my editor, Ian Straus, for having faith in the project. Having fully relaxed into cranky middle age, I will confess that I was initially dubious about working with Ian, who seemed to be half my age. This has, instead, been a humbling experience. With his astoundingly sharp intellectual insights, ability to see connections between material that I'd missed, and readiness to point out inconsistencies in my arguments and errors in the flow of my prose, Ian vastly improved this manuscript and helped me to think more clearly about what I wanted to say. I'd also like to thank Tracy Behar, publisher and editor-in-chief of Little, Brown Spark, for her enthusiasm for the book; my publicist, Stephanie Reddaway; marketing director Jess Chun; art director Lauren Harms for designing my favorite cover ever; production editor Ben Allen; and my copyeditor, Deri Reed, for assiduous cleaning up and checking of my prose.

Most of all, I would like to thank Thalia Wheatley. She read through the entire manuscript from earliest drafts, suggested wonderful examples, and forced me to confront serious problems with both my reasoning and my prose. She also attempted, with mixed success, to correct some of the science; remaining errors should be blamed solely upon me. Most of all, Thalia inspired me to think more deeply about intoxication, pleasure, and joy. This would have been a very different book without her.

BIBLIOGRAPHY

Abbey, Antonia, Tina Zawacki, and Philip Buck. (2005). "The effects of past sexual assault perpetration and alcohol consumption on men's reactions to women's mixed signals." *Journal of Social and Clinical Psychology*, 24, 129–155.

Abrams, Dominic, Tim Hopthrow, Lorne Hulbert, and Daniel Frings. (2006). "'Groupdrink'? The effect of alcohol on risk attraction among groups versus individuals." *Journal of Studies on Alcohol*, 67(4), 628–636.

Aiello, Daniel A., Andrew F. Jarosz, Patrick J. Cushen, and Jennifer Wiley. (2012). "Firing the executive: When an analytic approach to problem solving helps and hurts." *Journal of Problem Solving*, 4(2).

Allan, Sarah. (2007). "Erlitou and the formation of Chinese civilization: Toward a new paradigm." *Journal of Asian Studies*, 66(2), 461–496.

Allen, Robert. (1983). "Collective invention." *Journal of Economic Behavior & Organization*, 4(1), 1–24.

Amabile, Teresa M. (1979). "Effects of external evaluation on artistic creativity." *Journal of Personality and Social Psychology*, 37(2), 221–233.

Anderson, Thomas, Rotem Petranker, Daniel Rosenbaum, Cory R. Weissman, Le-Anh Dinh-Williams, Katrina Hui,...Norman A. S. Farb. (2019). "Microdosing psychedelics: Personality, mental health, and creativity differences in microdosers." *Psychopharmacology*, 236(2), 731–740.

Andrews, Michael. (2017). "Bar talk: Informal social interactions, alcohol prohibition, and invention." (Unpublished manuscript.)

Archer, Ruth, Cleo Alper, Laura Mack, Melanie Weedon, Manmohan Sharma, Andreas Sutter, and David Hosken. (Under consideration). "Alcohol alters female sexual behavior." *Cell Press*. https://papers.ssrn.com/sol3/papers.cfm?abstract_id=3378006.

Arranz-Otaegui, Amaia, Lara Gonzalez Carretero, Monica N. Ramsey, Dorian Q. Fuller, and Tobias Richter. (2018). "Archaeobotanical evidence reveals the origins of bread 14,400 years ago in northeastern Jordan." *Proceedings of the National Academy of Sciences*, 115(31), 7925–7930.

Arthur, J. W. (2014). "Beer through the ages." *Anthropology Now*, 6, 1–11.

Austin, Gregory. (1979). *Perspectives on the History of Psychoactive Substance Use*. Department of Health, Education, and Welfare, Public Health Service, Alcohol, Drug Abuse, and Mental Health Administration, National Institute on Drug Abuse.

Bahi, Amine. (2013). "Increased anxiety, voluntary alcohol consumption and ethanol-induced place preference in mice following chronic psychosocial stress." *Stress,* 16(4), 441–451.

Baier, Annette. (1994). *Moral Prejudices: Essays on Ethics.* Cambridge, MA: Harvard University Press.

Banaji, Mahzarin, and Claude M. Steele. (1989). "Alcohol and self-evaluation: Is a social cognition approach beneficial?" *Social Cognition* (Guilford Press Periodicals), 7(2), 137–151.

Banaji, Mahzarin, and Claude M. Steele. (1988). *Alcohol and Self-Inflation.* (Unpublished manuscript.) Yale University, New Haven, CT.

Barbaree, H. E., W. L. Marshall, E. Yates, and L. O. Lightfoot. (1983). "Alcohol intoxication and deviant sexual arousal in male social drinkers." *Behavior Research and Therapy,* 21(4), 365–373.

Barnard, Hans, Alek N. Dooley, Gregory Areshian, Boris Gasparyan, and Kym F. Faull. (2011). "Chemical evidence for wine production around 4000 BCE in the late Chalcolithic Near Eastern highlands." *Journal of Archaeological Science,* 38(5), 977–984.

Battcock, Mike, and Sue Azam-Ali. (1998). *Fermented Fruits and Vegetables: A Global Perspective.* Rome: FAO Agricultural Services.

Baudelaire, Charles. (1869). "Enivrez-vous," in *Le Spleen de Paris (Petits poèmes en prose).* Paris: Calmann-Lévy.

Baum-Baicker, Cynthia. (1985). "The psychological benefits of moderate alcohol consumption: A review of the literature." *Drug and Alcohol Dependence,* 15(4), 305–322.

Baumeister, Roy. (1991). *Escaping the Self: Alcoholism, Spirituality, Masochism, and Other Flights from the Burden of Selfhood.* New York: Basic Books.

Bègue, Laurent, Brad J. Bushman, Oulmann Zerhouni, Baptiste Subra, and Medhi Ourabah. (2013). "'Beauty is in the eye of the beer holder': People who think they are drunk also think they are attractive." *British Journal of Psychology,* 104(2), 225–234.

Beilock, Sian. (2010). *Choke: What the Secrets of the Brain Reveal About Getting It Right When You Have To.* New York: Free Press.

Berman, A. H., O. Molander, M. Tahir, P. Törnblom, M. Gajecki, K. Sinadinovic, and C. Andersson. (2020). "Reducing risky alcohol use via smartphone app skills training among adult internet help-seekers: A randomized pilot trial." *Front Psychiatry,* 11, 434.

Bershad, Anya K., Matthew G. Kirkpatrick, Jacob A. Seiden, and Harriet de Wit. (2015). "Effects of acute doses of prosocial drugs methamphetamine and alcohol on plasma oxytocin levels." *Journal of Clinical Psychopharmacology,* 35(3), 308–312.

Bettencourt, Luis, and Geoffrey West. (2010). "A unified theory of urban living." *Nature,* 467(7318), 912–913.

Blocker, Jack. (2006). "Kaleidoscope in Motion: Drinking in the United States, 1400–2000." In Mack Holt (Ed.), *Alcohol: A Social and Cultural History* (pp. 225–240). Oxford: Berg.

Bloom, Harold. (1992). *The American Religion: The Emergence of the Post-Christian Nation.* New York: Simon & Schuster.

Boone, R. Thomas, and Ross Buck. (2003). "Emotional expressivity and trustworthiness: The role of nonverbal behavior in the evolution of cooperation." *Journal of Nonverbal Behavior,* 27(3), 163–182.

Booth, C., and P. Hasking. (2009). "Social anxiety and alcohol consumption: The role of alcohol expectancies and reward sensitivity." *Addictive Behaviors,* 34(9), 730–736.

Borsari, B., and K. B. Carey. (2001). "Peer influences on college drinking: A review of the research." *Journal of Substance Abuse,* 13(4), 391–424.

Boseley, Sarah. (2018, August 23). "No Healthy Level of Alcohol Consumption, Says Major Study." *The Guardian.*

Bott, Elizabeth. (1987). "The Kava Ceremonial as a Dream Structure." In Mary Douglas (Ed.), *Constructive Drinking: Perspectives on Drink from Anthropology* (pp. 182–204). Cambridge: Cambridge University Press.

Boudreau, K. J., T. Brady, I. Ganguli, P. Gaule, E. Guinan, A. Hollenberg, and K. R. Lakhani. (2017). "A field experiment on search costs and the formation of scientific collaborations." *Review of Economics and Statistics,* 99(4), 565–576.

Bourguignon, Erika. (1973). *Religion, Altered States of Consciousness, and Social Change.* Columbus: Ohio State University Press.

Boyd, Robert, Peter Richerson, and Joseph Henrich. (2011). "The cultural niche: Why social learning is essential for human adaptation." *Proceedings of the National Academy of Sciences,* 108 (Supplement 2), 10918–10925.

Bradbury, Jack W., and Sandra L. Vehrencamp. (2000). "Economic models of animal communication." *Animal Behaviour,* 59(2), 259–268.

Braidwood, Robert J., Jonathan D. Sauer, Hans Helbaek, Paul C. Mangelsdorf, Hugh C. Cutler, Carleton S. Coon,...A. Leo Oppenheim. (1953). "Symposium: Did man once live by beer alone?" *American Anthropologist,* 55(4), 515–526.

Braun, Stephen. (1996). *Buzzed: The Science and Lore of Alcohol and Caffeine.* London: Penguin.

Bray, Tamara. (2009). "The Role of Chicha in Inca State Expansion: A Distributional Study of Inca Aríbalos." In Justin Jennings and Brenda Bowser (Eds.), *Drink, Power, and Society in the Andes* (pp. 108–132). Gainesville: University Press of Florida.

Britton, A., A. Singh-Manoux, and M. Marmot. (2004). "Alcohol consumption and cognitive function in the Whitehall II Study." *American Journal of Epidemiology,* 160(3), 240–247.

Brown, Stuart. (2009). *Play: How It Shapes the Brain, Opens the Imagination, and Invigorates the Soul.* New York: Penguin.

Bryant, G., and C. A. Aktipis. (2014). "The animal nature of spontaneous human laughter." *Evolution and Human Behavior,* 35, 327–335.

Bryner, Michelle. (2010, July 29). "How much alcohol is in my drink?" *Live Science.* Retrieved from https://www.livescience.com/32735-how-much-alcohol-is-in-my-drink.html.

Bushman, Brad J., and Harris M. Cooper. (1990). "Effects of alcohol on human aggression: An integrative research review." *Psychological Bulletin,* 107(3), 341–354.

Byers, John. (1997). *American Pronghorn: Social Adaptations and the Ghosts of Predators Past.* Chicago: University of Chicago Press.

Campos, Raquel, Fernanda Leon, and Ben McQuillin. (2018). "Lost in the storm: The academic collaborations that went missing in hurricane ISAAC." *Economic Journal,* 128(610), 995–1018.

Camus, Albert. (1955). *The Myth of Sisyphus and Other Essays* (Justin O'Brien, Trans.). New York: Vintage.

Capraro, Valerio, Jonathan Schulz, and David G. Rand. (2019). "Time pressure and honesty in a deception game." *Journal of Behavioral and Experimental Economics,* 79, 93–99.

Carhart-Harris, R. L., and K. J. Friston. (2019). "REBUS and the anarchic brain: Toward a unified model of the brain action of psychedelics." *Pharmacological Reviews,* 71, 316–344. 10.1124/pr.118.017160.

Carmody, S., J. Davis, S. Tadi, J. S. Sharp, R. K. Hunt, and J. Russ. (2018). "Evidence of tobacco from a Late Archaic smoking tube recovered from the Flint River site in southeastern North America." *Journal of Archaeological Science: Reports,* 21, 904–910.

Carod-Artal, F. J. (2015). "Hallucinogenic drugs in pre-Columbian Mesoamerican cultures." *Neurología* (English edition), 30(1), 42–49.

Carrigan, Matthew. (2020). "Hominoid Adaptation to Dietary Ethanol." In Kimberley Hockings and Robin Dunbar (Eds.), *Alcohol and Humans: A Long and Social Affair* (pp. 24–44). New York: Oxford University Press.

Carrigan, Matthew, Oleg Uryasev, Carole B. Frye, Blair L. Eckman, Candace R. Myers, Thomas D. Hurley, and Steven A. Benner. (2014). "Hominids adapted to metabolize ethanol long before human-directed fermentation." *Proceedings of the National Academy of Sciences,* 112(2), 458–463.

Centorrino, Samuele, Elodie Djemai, Astrid Hopfensitz, Manfred Milinski, and Paul Seabright. (2015). "Honest signaling in trust interactions: Smiles rated as genuine induce trust and signal higher earning opportunities." *Evolution and Human Behavior,* 36(1), 8–16.

Chan, Tak Kam. (2013). "From Conservatism to Romanticism: Wine and Prose-Writing from Pre-Qin to Jin." In Isaac Yue and Siufu Tang (Eds.), *Scribes of Gastronomy* (pp. 15–26). Hong Kong: Hong Kong University Press.

Chen, X., X. Wang, D. Yang, and Y. Chen. (2014). "The moderating effect of stimulus attractiveness on the effect of alcohol consumption on attractiveness ratings." *Alcohol and Alcoholism,* 49(5), 515–519.

Christakis, Nicholas. (2019). *Blueprint: The Evolutionary Origins of a Good Society.* New York: Little, Brown Spark.

Chrysikou, Evangelia. (2019). "Creativity in and out of (cognitive) control." *Current Opinion in Behavioral Sciences,* 27, 94–99.

Chrysikou, Evangelia, Roy H. Hamilton, H. Branch Coslett, Abhishek Datta, Marom Bikson, and Sharon L. Thompson-Schill. (2013). "Noninvasive transcranial direct current stimulation over the left prefrontal cortex facilitates cognitive flexibility in tool use." *Cognitive Neuroscience,* 4(2), 81–89.

Chrzan, Janet. (2013). *Alcohol: Social Drinking in Cultural Context.* New York: Routledge.

Cogsdill, E. J., A. T. Todorov, E. S. Spelke, and M. R. Banaji. (2014). "Inferring character from faces: A developmental study." *Psychological Science,* 25(5), 1132–1139.

BIBLIOGRAPHY

Collaborators, GDB Alcohol. (2018). "Alcohol use and burden for 195 countries and territories, 1990–2016: A systematic analysis for the Global Burden of Disease Study 2016." *The Lancet,* 392(10152), 1015–1035.

Collins, R. Lorraine, George A. Parks, and G. Alan Marlatt. (1985). "Social determinants of alcohol consumption: The effects of social interaction and model status on the self-administration of alcohol." *Journal of Consulting and Clinical Psychology,* 53(2), 189–200.

Courtwright, David. (2019). *The Age of Addiction: How Bad Habits Became Big Business.* Cambridge, MA: Harvard University Press.

Crockett, Molly J., Luke Clark, Marc D. Hauser, and Trevor W. Robbins. (2010). "Serotonin selectively influences moral judgment and behavior through effects on harm aversion." *Proceedings of the National Academy of Sciences,* 107(40), 17433.

Curry, Andrew. (2017, February). "Our 9,000-Year Love Affair with Booze." *National Geographic.*

Curtin, John, Christopher Patrick, Alan Lang, John Cacioppo, and Niels Birnbaumer. (2001). "Alcohol affects emotion through cognition." *Psychological Science,* 12(6), 527–531.

Dally, Joanna, Nathan Emery, and Nicola Clayton. (2006). "Food-caching Western scrub jays keep track of who was watching when." *Science,* 312(5780), 1662–1665.

Damasio, Antonio. (1994). *Descartes' Error: Emotion, Reason, and the Human Brain.* New York: G. P. Putnam's Sons.

Darwin, Charles. (1872/1998). *The Expression of Emotions in Man and Animals (With Introduction, Afterword and Commentaries by Paul Ekman).* New York: Oxford University Press.

Davidson, James. (2011). *Courtesans and Fishcakes: The Consuming Passions of Classical Athens.* Chicago: University of Chicago Press.

Dawkins, Richard. (1976/2006). *The Selfish Gene* (30th Anniversary Edition). Oxford: Oxford University Press.

Dawkins, Richard, John Richard Krebs, J. Maynard Smith, and Robin Holliday. (1979). "Arms races between and within species." *Proceedings of the Royal Society B: Biological Sciences,* 205(1161), 489–511.

de Bono, Edward. (1965). "Il cervello e il pensiero." In Angelo Majorana (Ed.), *Il cervello: organizzazione e funzioni* (pp. 203–208). Milan: Le Scienze.

Dean, Rosamund. (2017). *Mindful Drinking: How Cutting Down Can Change Your Life.* London: Orion Publishing Group.

DeCaro, Marci, Robin Thomas, Neil Albert, and Sian Beilock. (2011). "Choking under pressure: Multiple routes to skill failure." *Journal of Experimental Psychology,* 140(3), 390–406.

Dennis, Philip. (1979). "The Role of the Drunk in a Oaxacan Village." In Mac Marshall (Ed.), *Beliefs, Behaviors, and Alcoholic Beverages: A Cross-Cultural Survey* (pp. 54–63). Ann Arbor: University of Michigan Press.

DeSteno, D., C. Breazeal, R. H. Frank, D. Pizarro, J. Baumann, L. Dickens, and J. J. Lee. (2012). "Detecting the trustworthiness of novel partners in economic exchange." *Psychological Science,* 23(12), 1549–1556.

Devineni, A. V., and U. Heberlein. (2009). "Preferential ethanol consumption in Drosophila models features of addiction." *Current Biology*, 19(24), 2126–2132.

Dietler, Michael. (2006). "Alcohol: Anthropological/archaeological perspectives. *Annual Review of Anthropology*, 35, 229–249.

Dietler, Michael. (2020). "Alcohol as Embodied Material Culture: Anthropological Reflections of the Deep Entanglement of Humans and Alcohol." In Kimberley Hockings and Robin Dunbar (Eds.), *Alcohol and Humans: A Long and Social Affair* (pp. 115–129). New York: Oxford University Press.

Dietrich, Arne. (2003). "Functional neuroanatomy of altered states of consciousness: The transient hypofrontality hypothesis." *Consciousness and Cognition*, 12, 231–256.

Dietrich, Laura, Julia Meister, Oliver Dietrich, Jens Notroff, Janika Kiep, Julia Heeb,...Brigitta Schütt. (2019). "Cereal processing at Early Neolithic Göbekli Tepe, southeastern Turkey." *PLOS ONE*, 14(5), e0215214.

Dietrich, Oliver, and Laura Dietrich. (2020). "Rituals and Feasting as Incentives for Cooperative Action at Early Neolithic Göbekli Tepe." In Kimberley Hockings and Robin Dunbar (Eds.), *Alcohol and Humans: A Long and Social Affair* (pp. 93–114). New York: Oxford University Press.

Dietrich, Oliver, Manfred Heun, Jens Notroff, Klaus Schmidt, and Martin Zarnkow. (2012). "The role of cult and feasting in the emergence of Neolithic communities. New evidence from Göbekli Tepe, south-eastern Turkey." *Antiquity*, 86(333), 674–695.

Dijk, Corine, Bryan Koenig, Tim Ketelaar, and Peter de Jong. (2011). "Saved by the blush: Being trusted despite defecting." *Emotion*, 11(2), 313–319.

Dineley, Merryn. (2004). *Barley, Malt and Ale in the Neolithic Near East, 10,000–50,000*. Oxford: BAR Publishing.

Djos, Matts. (2010). *Writing Under the Influence: Alcoholism and the Alcoholic Perception from Hemingway to Berryman*. London: Palgrave Macmillan.

Doblin, Rick. (1991). "Pahnke's 'Good Friday Experiment': A long-term follow-up and methodological critique." *Journal of Transpersonal Psychology*, 23(1), 1–28.

Dolder, Patrick, Friederike Holze, Evangelia Liakoni, Samuel Harder, Yasmin Schmid, and Matthias Liechti. (2016). "Alcohol acutely enhances decoding of positive emotions and emotional concern for positive stimuli and facilitates the viewing of sexual images." *Psychopharmacology*, 234, 41–51.

Dominguez-Clave, E., J. Soler, M. Elices, J. C. Pascual, E. Alvarez, M. de la Fuente Revenga,...J. Riba. (2016). "Ayahuasca: Pharmacology, neuroscience and therapeutic potential." *Brain Research Bulletin*, 126(Part 1), 89–101.

Dominy, Nathaniel J. (2015). "Ferment in the family tree." *Proceedings of the National Academy of Sciences*, 112(2), 308.

Doniger O'Flaherty, Wendy. (1968). "The Post-Vedic History of the Soma Plant." In R. Gordon Wasson (Ed.), *Soma: Divine Mushroom of Immortality* (pp. 95–147). New York: Harcourt Brace.

Doughty, Paul. (1979). "The Social Uses of Alcoholic Beverages in a Peruvian Community." In Mac Marshall (Ed.), *Beliefs, Behaviors, and Alcoholic Beverages: A Cross-Cultural Survey* (pp. 64–81). Ann Arbor: University of Michigan Press.

Douglas, Mary (Ed.). (1987). *Constructive Drinking: Perspectives on Drink from Anthropology*. Cambridge: Cambridge University Press.

Dry, Matthew J., Nicholas R. Burns, Ted Nettelbeck, Aaron L. Farquharson, and Jason M. White. (2012). "Dose-related effects of alcohol on cognitive functioning." *PLOS ONE,* 7(11), e50977–e50977.

Dudley, Robert. (2014). *The Drunken Monkey: Why We Drink and Abuse Alcohol.* Berkeley: University of California Press.

Dudley, Robert. (2020). "The Natural Biology of Dietary Ethanol, and Its Implications for Primate Evolution." In Kimberley Hockings and Robin Dunbar (Eds.), *Alcohol and Humans: A Long and Social Affair* (pp. 9–23). New York: Oxford University Press.

Duke, Guy. (2010). "Continuity, Cultural Dynamics, and Alcohol: The Reinterpretation of Identity Through Chicha in the Andes." In L. Amundsen-Meyer, N. Engel, and S. Pickering (Eds.), *Identity Crisis: Archaeological Perspectives on Social Identity* (pp. 263–272). Calgary: University of Calgary Press.

Dunbar, Robin. (2014). "How conversations around campfires came to be." *Proceedings of the National Academy of Sciences,* 111(39), 14013–14014.

Dunbar, Robin. (2017). "Breaking bread: The functions of social eating." *Adaptive Human Behavior and Physiology,* 3(3), 198–211.

Dunbar, Robin. (2018, August 9). "Why Drink Is the Secret to Humanity's Success." *Financial Times.*

Dunbar, Robin, and Kimberley Hockings. (2020). "The Puzzle of Alcohol Consumption." In Kimberley Hockings and Robin Dunbar (Eds.), *Alcohol and Humans: A Long and Social Affair* (pp. 1–8). New York: Oxford University Press.

Dunbar, Robin, Jacques Launay, Rafael Wlodarski, Cole Robertson, Eiluned Pearce, James Carney, and Pádraig MacCarron. (2016). "Functional benefits of (modest) alcohol consumption." *Adaptive Human Behavior and Physiology,* 3(2), 118–133.

Durkheim, Émile. (1915/1965). *The Elementary Forms of the Religious Life* (Joseph Ward Swain, Trans.). New York: George Allen and Unwin Ltd.

Dutton, H. I. (1984). *The Patent System and Inventive Activity During the Industrial Revolution, 1750–1852.* Manchester: Manchester University Press.

Earle, Rebecca. (2014). "Indians and drunkenness in Spanish America." *Past and Present,* 222(Supplement 9), 81–99.

Easdon, C., A. Izenberg, M. L. Armilio, H. Yu, and C. Alain. (2005). "Alcohol consumption impairs stimulus- and error-related processing during a Go/No-Go Task." *Brain Research. Cognitive Brain Research,* 25(3), 873–883.

Edwards, Griffith. (2000). *Alcohol: The World's Favorite Drug.* New York: Thomas Dunne Books.

Ehrenreich, Barbara. (2007). *Dancing in the Streets: A History of Collective Joy.* New York: Metropolitan Books.

Ekman, Paul. (2003). *Emotions Revealed: Recognizing Faces and Feelings to Improve Communication and Emotional Life.* New York: Times Books.

Ekman, Paul. (2006). *Darwin and Facial Expression: A Century of Research in Review.* Los Altos, CA: Malor Books.

Ekman, Paul, and Wallace V. Friesen. (1982). "Felt, false, and miserable smiles." *Journal of Nonverbal Behavior,* 6(4), 238–252.

Ekman, Paul, and M. O'Sullivan. (1991). "Who can catch a liar? "*American Psychologist,* 46, 913–920.

Eliade, Mircea. (1964). *Shamanism: Archaic Techniques of Ecstasy* (Revised and Enlarged Edition). New York: Bollingen Foundation.

Emery, Nathan, and Nicola Clayton. (2004). "The mentality of crows: Convergent evolution of intelligence in corvids and apes." *Science,* 306(5703), 1903–1907.

Eno, Robert. (2009). "Shang State Religion and the Pantheon of the Oracle Texts." In John Lagerwey and Marc Kalinowski (Eds.), *Early Chinese Religion: Part One: Shang Through Han (1250 BC–22 AD)* (pp. 41–102). Leiden: Brill.

Enright, Michael. (1996). *Lady with a Mead Cup: Ritual, Prophecy, and Lordship in the European Warband from La Tène to the Viking Age.* Portland, OR: Four Courts Press.

Fairbairn, C. E., M. Sayette, O. Aelen, and A. Frigessi. (2015). "Alcohol and emotional contagion: An examination of the spreading of smiles in male and female drinking groups." *Clinical Psychological Science,* 3(5), 686–701.

Fairbairn, C. E., and M. Testa. (2016). "Relationship quality and alcohol-related social reinforcement during couples interaction." *Clinical Psychological Science,* 5(1), 74–84.

Farris, Coreen, Teresa A. Treat, and Richard J. Viken. (2010). "Alcohol alters men's perceptual and decisional processing of women's sexual interest." *Journal of Abnormal Psychology,* 119(2), 427–432.

Fatur, K. (2019). "Sagas of the Solanaceae: Speculative ethnobotanical perspectives on the Norse berserkers." *Journal of Ethnopharmacology,* 244, 112151.

Fay, Justin C., and Joseph A. Benavides. (2005). "Evidence for domesticated and wild populations of saccharomyces cerevisiae." *PLOS Genetics,* 1(1), e5.

Feinberg, Matthew, Robb Willer, and Dacher Kaltner. (2011). "Flustered and faithful: Embarrassment as a signal of prosociality." *Journal of Personality and Social Psychology,* 102(1), 81–97.

Fernandez, James. (1972). "*Tabernanthe Iboga*: Narcotic Ecstasis and the Work of the Ancestors." In Peter Furst (Ed.), *Flesh of the Gods: The Ritual Use of Hallucinogens* (pp. 237–260). New York: Praeger.

Fertel, Randy. (2015). *A Taste for Chaos: The Art of Literary Improvisation.* New Orleans: Spring Journal Books.

Florida, Richard. (2002). *The Rise of the Creative Class: And How It's Transforming Work, Leisure, Community and Everyday Life.* New York: Basic Books.

Forstmann, M., D. A. Yudkin, A. M. B. Prosser, S. M. Heller, and M. J. Crockett. (2020). "Transformative experience and social connectedness mediate the mood-enhancing effects of psychedelic use in naturalistic settings." *Proceedings of the National Academy of Sciences,* 117(5), 2338–2346.

Forsyth, Mark. (2017). *A Short History of Drunkenness.* New York: Viking.

Fox, K. C. R., M. Muthukrishna, and S. Shultz. (2017). "The social and cultural roots of whale and dolphin brains." *Nature Ecology and Evolution,* 1(11), 1699–1705.

Frank, Mark G., and Paul Ekman. (1997). "The ability to detect deceit generalizes across different types of high-stake lies." *Journal of Personality and Social Psychology,* 72(6), 1429–1439.

Frank, Robert. (1988). *Passions Within Reason: The Strategic Role of the Emotions.* New York: W. W. Norton & Company.

Frank, Robert. (2001). "Cooperation Through Emotional Commitment." In

BIBLIOGRAPHY

Randolph M. Nesse (Ed.), *Evolution and the Capacity for Commitment* (pp. 57–76). New York: Russell Sage Foundation.

Frank, Robert, T. Gilovich, and D. T. Regan. (1993). "The evolution of one-shot cooperation: An experiment." *Ethology and Sociobiology*, 14, 247–256.

Frey, Carl Benedikt. (2020, July 8). "The Great Innovation Deceleration: Our response to the Covid-19 pandemic could damage the world's collective brain." *MIT Sloan Management Review*.

Frings, Daniel, Tim Hopthrow, Dominic Abrams, Lorne Hulbert, and Roberto Gutierrez. (2008). "'Groupdrink': The effects of alcohol and group process on vigilance errors." Group Dynamics: Theory, Research, and Practice, 12(3), 179–190.

Fromme, Kim, G. Alan Marlatt, John S. Baer, and Daniel R. Kivlahan. (1994). "The alcohol skills training program: A group intervention for young adult drinkers." *Journal of Substance Abuse Treatment*, 11(2), 143–154.

Frye, Richard. (2005). *Ibn Fadlan's Journey to Russia*. Princeton, NJ: Markus Wiener.

Fuller, Robert. (1995). "Wine, symbolic boundary setting, and American religious communities." *Journal of the American Academy of Religion*, 63(3), 497–517.

Fuller, Robert. (2000). *Stairways to Heaven: Drugs in American Religious History*. Boulder, CO: Westview Press.

Furst, Peter. (1972). "To Find Our Life: Peyote Among the Huichol Indians of Mexico." In Peter Furst (Ed.), *Flesh of the Gods: The Ritual Use of Hallucinogens* (pp. 136–184). New York: Praeger.

Gable, Shelly L., Elizabeth A. Hopper, and Jonathan W. Schooler. (2019). "When the muses strike: Creative ideas of physicists and writers routinely occur during mind wandering." *Psychological Science*, 30(3), 396–404.

Garvey, Pauline. (2005). "Drunk and (Dis)orderly: Norwegian Drinking Parties in the Home." In Thomas Wilson (Ed.), *Drinking Cultures: Alcohol and Identity* (pp. 87–106). Oxford: Berg.

Gately, Iain. (2008). *Drink: A Cultural History of Alcohol*. New York: Gotham Books.

Gefou-Madianou, Dimitra (Ed.). (1992). *Alcohol, Gender and Culture*. London: Routledge.

George, Andrew. (2003). *The Epic of Gilgamesh*. New York: Penguin.

George, W. H., and S. A. Stoner. (2000). "Understanding acute alcohol effects on sexual behavior." *Annual Review of Sex Research*, 11, 92–124.

Gerbault, Pascale, Anke Liebert, Yuval Itan, Adam Powell, Mathias Currat, Joachim Burger,...Mark G. Thomas. (2011). "Evolution of lactase persistence: An example of human niche construction." *Philosophical Transactions of the Royal B: Biological Sciences*, 366(1566), 863–877.

Getting to Zero Alcohol-Impaired Driving Fatalities: A Comprehensive Approach to a Persistent Problem. (2018). Washington, DC: National Academies Press.

Giancola, Peter R. (2002). "The influence of trait anger on the alcohol-aggression relation in men and women." *Alcoholism: Clinical and Experimental Research*, 26(9), 1350–1358.

Gianoulakis, Christina. (2004). "Endogenous opioids and addition to alcohol and drugs of abuse." *Current Topics in Medical Chemistry*, 4, 39–50.

Gibbons, Ann. (2013, February 16). "Human Evolution: Gain Came with Pain." *Science News*.

Gladwell, Malcolm. (2010, February 8). "Drinking Games: How Much People Drink May Matter Less Than How They Drink It." *The New Yorker*.

Glassner, Barry. (1991). "Jewish Sobriety." In David Pittman and Helene Raskin White (Eds.), *Society, Culture, and Drinking Patterns Reexamined* (pp. 311–326). New Brunswick, NJ: Rutgers Center of Alcohol Studies.

Gochman, Samuel R., Michael B. Brown, and Nathaniel J. Dominy. (2016). "Alcohol discrimination and preferences in two species of nectar-feeding primate." *Royal Society Open Science*, 3(7), 160217.

Goldman, D., and M. A. Enoch. (1990). "Genetic epidemiology of ethanol metabolic enzymes: A role for selection." *World Review of Nutrition and Dietetics*, 63, 143–160.

Gopnik, Alison. (2009). *The Philosophical Baby: What Children's Minds Tell Us About Truth, Love, and the Meaning of Life*. New York: Farrar, Straus and Giroux.

Gopnik, Alison, S. O'Grady, C. G. Lucas, T. L. Griffiths, A. Wente, S. Bridgers, ... R. E. Dahl. (2017). "Changes in cognitive flexibility and hypothesis search across human life history from childhood to adolescence to adulthood." *Proceedings of the National Academy of Sciences*, 114(30), 7892–7899.

Grant B. F., R. B. Goldstein, T. D. Saha, S. P. Chou, J. Jung, H. Zhang, R. P. Pickering, W. J. Ruan, S. M. Smith, B. Huang, and D. S. Hasin. (2015). "Epidemiology of DSM-5 alcohol use disorder: Results from the national epidemiologic survey on alcohol and related conditions III." *JAMA Psychiatry*, 72(8), 757–766.

Griffiths, R. R., M. W. Johnson, W. A. Richards, B. D. Richards, U. McCann, and R. Jesse. (2011). "Psilocybin occasioned mystical-type experiences: Immediate and persisting dose-related effects." *Psychopharmacology*, 218(4), 649–665.

Guasch-Jané, Maria Rosa. (2008). *Wine in Ancient Egypt: A Cultural and Analytical Study*. Oxford: Archaeopress.

Guerra-Doce, Elisa. (2014). "The origins of inebriation: Archaeological evidence of the consumption of fermented beverages and drugs in prehistoric Eurasia." *Journal of Archaeological Method and Theory*, 22(3), 751–782.

Guerra-Doce, Elisa. (2020). "The Earliest Toasts: Archeological Evidence for the Social and Cultural Construction of Alcohol in Prehistoric Europe." In Kimberley Hockings and Robin Dunbar (Eds.), *Alcohol and Humans: A Long and Social Affair* (pp. 60–80). New York: Oxford University Press.

Haarmann, Henk, Timothy George, Alexei Smaliy, and Joseph Dien. (2012). "Remote associates test and alpha brain waves." *Journal of Problem Solving*, 4(2).

Hagen, E. H., C. J. Roulette, and R. J. Sullivan. (2013). "Explaining human recreational use of 'pesticides': The neurotoxin regulation model of substance use vs. the hijack model and implications for age and sex differences in drug consumption." *Front Psychiatry*, 4, 142.

Hagen, E., and Shannon Tushingham. (2019). "The Prehistory of Psychoactive Drug Use." In Tracy Henley, Matthew Rossano, and Edward Kardas (Eds.), *Cognitive Archaeology: Psychology in Prehistory*. New York: Routledge.

Haidt, Jonathan. (2001). "The emotional dog and its rational tail: A social intuitionist approach to moral judgment." *Psychological Review*, 108(4), 814–834.

Haidt, Jonathan, J. Patrick Seder, and Selin Kesebir. (2008). "Hive psychology, happiness, and public policy." *Journal of Legal Studies*, 37, 133–156.

Hall, Timothy. (2005). "Pivo at the Heart of Europe: Beer-Drinking and Czech

Identity." In Thomas Wilson (Ed.), *Drinking Cultures: Alcohol and Identity* (pp. 65–86). Oxford: Berg.

Han, Y., S. Gu, H. Oota, M. V. Osier, A. J. Pakstis, W. C. Speed,...K. K. Kidd. (2007). "Evidence of positive selection on a class I ADH locus." *American Journal of Human Genetics*, 80(3), 441–456.

Harkins, Stephen G. (2006). "Mere effort as the mediator of the evaluation-performance relationship." *Journal of Personality and Social Psychology*, 91(3), 436–455.

Hart, H. H. (1930). "Personality factors in alcoholism." *Archives of Neurology and Psychiatry*, 24, 116–134.

Hauert, C., S. De Monte, J. Hofbauer, and K. Sigmund. (2002). "Volunteering as Red Queen mechanism for cooperation in public goods games." *Science*, 296(5570), 1129–1132.

Hayden, Brian. (1987). "Alliances and ritual ecstasy: Human responses to resource stress." *Journal for the Scientific Study of Religion*, 26(1), 81–91.

Hayden, Brian, Neil Canuel, and Jennifer Shanse. (2013). "What was brewing in the Natufian? An archaeological assessment of brewing technology in the Epipaleolithic." *Journal of Archaeological Method and Theory*, 20(1), 102–150.

Heath, Dwight. (1958). "Drinking patterns of the Bolivian Camba." *Quarterly Journal of Studies on Alcohol*, 19(3), 491–508.

Heath, Dwight. (1976). "Anthropological Perspectives on Alcohol: An Historical Review." In Michael Everett, Jack Waddell, and Dwight Heath (Eds.), *Cross-Cultural Approaches to the Study of Alcohol: An Interdisciplinary Perspective*. The Hague: Mouton Publishers.

Heath, Dwight. (1987). "A Decade of Development in the Anthropological Study of Alcohol Use, 1970–1980." In Mary Douglas (Ed.), *Constructive Drinking: Perspectives on Drink from Anthropology* (pp. 16–69). Cambridge: Cambridge University Press.

Heath, Dwight. (1990). "Anthropological and Sociocultural Perspectives on Alcohol as a Reinforcer." In W. Miles Cox (Ed.), *Why People Drink: Parameters of Alcohol as a Reinforcer* (pp. 263–290). New York: Gardner Press.

Heath, Dwight (1994). "Agricultural changes and drinking among the Bolivian Camba: A longitudinal view of the aftermath of a revolution." *Human Organization*, 53(4), 357–361.

Heath, Dwight. (2000). *Drinking Occasions: Comparative Perspectives on Alcohol and Culture*. New York: Routledge.

Heaton, R., G. Chelune, J. Talley, G. Kay, and G. Curtiss. (1993). *Wisconsin Card Sorting Test Manual: Revised and Expanded*. Lutz, FL: Psychological Assessment Resources.

Heberlein, Ulrike, Fred W. Wolf, Adrian Rothenfluh, and Douglas J. Guarnieri. (2004). "Molecular genetic analysis of ethanol intoxication in drosophila melanogaster1." *Integrative and Comparative Biology*, 44(4), 269–274.

Heidt, Amanda. (2020, June 8). "Like Humans, These Big-Brained Birds May Owe Their Smarts to Long Childhoods." *Science News*.

Heinrich, Bernd. (1995). "An experimental investigation of insight in common ravens (Corvus corax)." *The Auk*, 112(4), 994–1003.

Henrich, Joseph. (2015). *The Secret of Our Success: How Culture Is Driving Human*

Evolution, Domesticating Our Species, and Making Us Smarter. Princeton, NJ: Princeton University Press.

Henrich, Joseph, and Richard McElreath. (2007). "Dual Inheritance Theory: The Evolution of Human Cultural Capacities and Cultural Evolution." In Robin Dunbar and Louise Barrett (Eds.), *Oxford Handbook of Evolutionary Psychology* (pp. 555–570). Oxford: Oxford University Press.

Hertenstein, Elisabeth, Elena Waibel, Lukas Frase, Dieter Riemann, Bernd Feige, Michael Nitsche,...Christoph Nissen. (2019). "Modulation of creativity by transcranial direct current stimulation." *Brain Stimulation,* 12(5), 1213–1221.

Hirsch, Jacob, Adam Galinsky, and Chen-bo Zhong. (2011). "Drunk, powerful, and in the dark: How general processes of disinhibition produce both prosocial and antisocial behavior." *Perspectives on Psychological Science,* 6(5), 415–427.

Hockings, Kimberley, and Robin Dunbar (Eds.). (2020). *Alcohol and Humans: A Long and Social Affair.* New York: Oxford University Press.

Hockings, Kimberley, Miho Ito, and Gen Yamakoshi. (2020). "The Importance of Raffia Palm Wine to Coexisting Humans and Chimpanzees." In Kimberley Hockings and Robin Dunbar (Eds.), *Alcohol and Humans: A Long and Social Affair* (pp. 45–59). New York: Oxford University Press.

Hogan, Emma. (2017, August 1). "Turn On, Tune In, Drop by the Office." *The Economist 1843.*

Holtzman, Jon. (2001). "The food of elders, the 'ration' of women: Brewing, gender, and domestic processes among the Samburu of Northern Kenya." *American Anthropologist,* 103(4), 1041–1058.

Horton, Donald. (1943). "The functions of alcohol in primitive societies: A cross-cultural study." *Quarterly Journal of Studies on Alcohol,* 4, 199–320.

Hrdy, Sarah Blaffer. (2009). *Mothers and Others: The Evolutionary Origins of Mutual Understanding.* Cambridge, MA: Belknap Press.

Huizinga, Johan. (1955). *Homo Ludens: A Study of the Play Element in Culture.* Boston: Beacon Press.

Hull, Jay G., Robert W. Levenson, Richard David Young, and Kenneth J. Sher. (1983). "Self-awareness-reducing effects of alcohol consumption." *Journal of Personality and Social Psychology,* 44(3), 461–473.

Hull, Jay G., and Laurie B. Slone. (2004). "Alcohol and Self-Regulation." In Roy F. Baumeister and Kathleen D. Vohs (Eds.), *Handbook of Self-Regulation: Research, Theory, and Applications* (pp. 466–491). New York: Guilford Press.

Hunt, Tristan. (2009). *The Frock-Coated Communist: The Revolutionary Life of Friedrich Engels.* London: Penguin.

Hurley, C., and Mark G. Frank. (2011). "Executing facial control during deception situations." *Journal of Nonverbal Behavior,* 35, 119–131.

Huxley, Aldous. (1954/2009). *The Doors of Perception.* New York: HarperCollins.

Hyman, S. E. (2005). "Addiction: A disease of learning and memory." *American Journal of Psychiatry,* 168(8), 1414–1422.

Ilardo, Melissa A., Ida Moltke, Thorfinn S. Korneliussen, Jade Cheng, Aaron J. Stern, Fernando Racimo,...Eske Willerslev. (2018). "Physiological and genetic adaptations to diving in sea nomads." *Cell,* 173(3), 569–580.e515.

Ing, Michael. (In preparation). *What Remains: Grief and Resilience in the Thought of Tao Yuanming.*

James, William. (1902/1961). *The Varieties of Religious Experience: A Study in Human Nature.* New York: Collier Books.

Jarosz, A. F., G. J. Colflesh, and J. Wiley. (2012). "Uncorking the muse: Alcohol intoxication facilitates creative problem solving." *Consciousness and Cognition,* 21(1), 487–493.

Jennings, Justin, and Brenda Bowser. (2009). "Drink, Power, and Society in the Andes: An Introduction." In Justin Jennings and Brenda Bowser (Eds.), *Drink, Power, and Society in the Andes* (pp. 1–27). Gainesville: University Press of Florida.

Joe-Laidler, Karen, Geoffrey Hunt, and Molly Moloney. (2014). "'Tuned Out or Tuned In': Spirituality and Youth Drug Use in Global Times." In Phil Withington and Angela McShane (Eds.), *Cultures of Intoxication, Past and Present* (Vol. 222, pp. 61–80). Oxford: Oxford University Press.

Joffe, Alexander. (1998). "Alcohol and social complexity in Ancient Western Asia." *Current Anthropology,* 39(3), 297–322.

Katz, Solomon, and Mary Voight. (1986). "Bread and beer." *Expedition,* 28(2), 23–35.

Khazan, Olga. (2020, January 14). "America's Favorite Poison: Whatever Happened to the Anti-Alcohol Movement?" *The Atlantic.*

Kirchner, T. R., M. A. Sayette, J. F. Cohn, R. L. Moreland, and J. M. Levine. (2006). "Effects of alcohol on group formation among male social drinkers." *Journal of Studies on Alcohol and Drugs,* 67(5), 785–793.

Kirkby, Diane. (2006). "Drinking 'The Good Life': Australia c.1880–1980." In Mack Holt (Ed.), *Alcohol: A Social and Cultural History* (pp. 203–224). Oxford: Berg.

Klatsky, Arthur L. (2004). "Alcohol and cardiovascular health." *Integrative and Comparative Biology,* 44(4), 324–328.

Kline, Michelle A., and Robert Boyd. (2010). "Population size predicts technological complexity in Oceania." *Proceedings of the Royal Society B: Biological Sciences,* 277(1693), 2559–2564.

Koenig, Debbie. (2019, December 21). "Not Just January: Alcohol Abstinence Turns Trendy." *WebMD.* https://www.webmd.com/mental-health/addiction/news/20191231/not-just-january-alcohol-abstinence-turns-trendy.

Kometer, M., T. Pokorny, E. Seifritz, and F. X. Vollenweider. (2015). "Psilocybin-induced spiritual experiences and insightfulness are associated with synchronization of neuronal oscillations." *Psychopharmacology,* 232(19), 3663–3676.

Koob, George F. (2003). "Alcoholism: Allostasis and Beyond." *Alcoholism: Clinical and Experimental Research,* 27(2), 232–243.

Koob, George F., and Michel Le Moal. (2008). "Addiction and the brain antireward system." *Annual Review of Psychology,* 59, 29–53.

Krumhuber, Eva, Antony S. R. Manstead, Darren Cosker, Dave Marshall, Paul Rosin, and Arvid Kappas. (2007). "Facial dynamics as indicators of trustworthiness and cooperative behavior." *Emotion,* 7(4), 730–735.

Kuhn, Cynthia, and Scott Swartzwelder. (1998). *Buzzed: The Straight Facts About the Most Used and Abused Drugs from Alcohol to Ecstasy.* New York: Penguin.

Kwong, Charles. (2013). "Making Poetry with Alcohol: Wine Consumption in Tao Qian, Li Bai and Su Shi." In Isaac Yue and Siufu Tang (Eds.), *Scribes of Gastronomy* (pp. 45–67). Hong Kong: Hong Kong University Press.

Laland, Kevin, John Odling-Smee, and Marcus Feldman. (2000). "Niche construction, biological evolution, and cultural change." *Behavioral and Brain Sciences,* 23(1), 131–175.

Lalander, Philip. (1997). "Beyond everyday order: Breaking away with alcohol." *Nordic Studies on Alcohol and Drugs,* 14(1_supplement), 33–42.

Lane, Scott, Don Cherek, Cynthia Pietras, and Oleg Tcheremissine. (2004). "Alcohol effects on human risk taking." *Psychopharmacology,* 172(1), 68–77.

Lang, I., R. B. Wallace, F. A. Huppert, and D. Melzer. (2007). "Moderate alcohol consumption in older adults is associated with better cognition and well-being than abstinence." *Age and Ageing,* 36(3), 256–261.

Larimer, Mary, and Jessica Cronce. (2007). "Identification, prevention, and treatment revisited: Individual-focused college drinking prevention strategies 1999–2006." *Addictive Behaviors,* 32(11), 2439–2468.

Leary, Mark. (2004). *The Curse of the Self: Self-Awareness, Egotism, and the Quality of Human Life.* New York: Oxford University Press.

Leary, Timothy. (2008). *Leary on Drugs: Writings and Lectures from Timothy Leary (1970–1996).* San Francisco: Re/Search Publications.

Lebot, Vincent, Lamont Lindstrom, and Mark Merlin. (1992). *Kava: The Pacific Drug.* New Haven, CT: Yale University Press.

Lee, H. G., Y. C. Kim, J. S. Dunning, and K. A. Han. (2008). "Recurring ethanol exposure induces disinhibited courtship in Drosophila." *PLOS ONE,* 3(1), e1391.

Lemmert, Edwin. (1991). "Alcohol, Values and Social Control." In David Pittman and Helene Raskin White (Eds.), *Society, Culture, and Drinking Patterns Reexamined* (pp. 681–701). New Brunswick, NJ: Rutgers Center of Alcohol Studies.

Levenson, Robert W., Kenneth J. Sher, Linda M. Grossman, Joseph Newman, and David B. Newlin. (1980). "Alcohol and stress response dampening: Pharmacological effects, expectancy, and tension reduction." *Journal of Abnormal Psychology,* 89(4), 528–538.

Levine, E. E., A. Barasch, D. Rand, J. Z. Berman, and D. A. Small. (2018). "Signaling emotion and reason in cooperation." *Journal of Experimental Psychology: General,* 147(5), 702–719.

Levitt, A., and M. Lynne Cooper. (2010). "Daily alcohol use and romantic relationship functioning: Evidence of bidirectional, gender-, and context-specific effects." *Personality and Social Psychology Bulletin,* 36(12), 1706–1722.

Levitt, A., J. L. Derrick, and M. Testa. (2014). "Relationship-specific alcohol expectancies and gender moderate the effects of relationship drinking contexts on daily relationship functioning." *Journal of Studies on Alcohol and Drugs,* 75(2), 269–278.

Lietava, Jan. (1992). "Medicinal plants in a Middle Paleolithic grave Shanidar IV?" *Journal of Ethnopharmacology,* 35(3), 263–266.

Limb, Charles J., and Allen R. Braun. (2008). "Neural substrates of spontaneous musical performance: An fMRI study of jazz improvisation." *PLOS ONE,* 3(2), e1679.

Long, Tengwen, Mayke Wagner, Dieter Demske, Christian Leipe, and Pavel E. Tarasov. (2016). "Cannabis in Eurasia: Origin of human use and Bronze Age trans-continental connections." *Vegetation History and Archaeobotany,* 26(2), 245–258.

Lu, Dongsheng, Haiyi Lou, Kai Yuan, Xiaoji Wang, Yuchen Wang, Chao Zhang,...Shuhua Xu. (2016). "Ancestral origins and genetic history of Tibetan Highlanders." *American Journal of Human Genetics,* 99(3), 580–594.

Lutz, H. F. (1922). *Viticulture and Brewing in the Ancient Orient.* Leipzig: J. C. Hinrichs.

Luyster, Robert. (2001). "Nietzsche/Dionysus: Ecstasy, heroism, and the monstrous." *Journal of Nietzsche Studies,* 21, 1–26.

Lyvers, Michael. (2000). "'Loss of control' in alcoholism and drug addiction: A neuroscientific interpretation." *Experimental and Clinical Psychopharmacology,* 8(2), 225–245.

Lyvers, Michael, Emma Cholakians, Megan Puorro, and Shanti Sundram. (2011). "Beer goggles: Blood alcohol concentration in relation to attractiveness ratings for unfamiliar opposite sex faces in naturalistic settings." *Journal of Social Psychology,* 151(1), 105–112.

Lyvers, Michael, N. Mathieson, and M. S. Edwards. (2015). "Blood alcohol concentration is negatively associated with gambling money won on the Iowa gambling task in naturalistic settings after controlling for trait impulsivity and alcohol tolerance." *Addictive Behaviors,* 41, 129–135.

Lyvers, Michael, and Juliette Tobias-Webb. (2010). "Effects of acute alcohol consumption on executive cognitive functioning in naturalistic settings." *Addictive Behaviors,* 35(11), 1021–1028.

Ma, Chengyuan (Ed.). (2012). *Shanghai Bowuguan Cang Zhanguo Chu Zhushu IX* 上海博物館藏戰國楚竹書(九). Shanghai: Shanghai Guji.

MacAndrew, Craig, and Robert B. Edgerton. (1969). *Drunken Comportment: A Social Explanation.* Chicago: Aldine.

Machin, A. J., and R. I. M. Dunbar. (2011). "The brain opioid theory of social attachment: A review of the evidence." *Behavior,* 148, 985–1025.

MacLean, Katherine, Matthew Johnson, and Roland Griffiths. (2011). "Mystical experiences occasioned by the hallucinogen psilocybin lead to increases in the personality domain of openness." *Journal of Psychopharmacology,* 25(11), 1453–1461.

Madsen, William, and Claudia Madsen. (1979). "The Cultural Structure of Mexican Drinking Behavior." In Mac Marshall (Ed.), *Beliefs, Behaviors, and Alcoholic Beverages: A Cross-Cultural Survey* (pp. 38–53). Ann Arbor: University of Michigan Press.

Mäkelä, Klaus. (1983). "The uses of alcohol and their cultural regulation." *Acta Sociologica,* 26(1), 21–31.

Mandelbaum, David. (1965). "Alcohol and culture." *Current Anthropology,* 6(3), 281–288 + 289–293.

Manthey, Jakob, Kevin D. Shield, Margaret Rylett, Omer S. M. Hasan, Charlotte Probst, and Jürgen Rehm. (2019). "Global alcohol exposure between 1990 and 2017 and forecasts until 2030: A modelling study." *The Lancet,* 393(10190), 2493–2502.

Marino, L. (2017). "Thinking chickens: A review of cognition, emotion, and behavior in the domestic chicken." *Animal Cognition,* 20(2), 127–147.

Markoff, John. (2005). *What the Dormouse Said: How the Sixties Counterculture Shaped the Personal Computer Industry.* New York: Viking.

Markos, A. R. (2005). "Alcohol and sexual behaviour." *International Journal of STD and AIDS,* 16(2), 123–127.

Marlatt, Alan, Mary Larimer, and Katie Witkiewitz (Eds.). (2012). *Harm Reduction, Second Edition: Pragmatic Strategies for Managing High-Risk Behaviors.* New York: Guilford Press.

Mars, Gerald. (1987). "Longshore Drinking, Economic Security and Union Politics in Newfoundland." In Mary Douglas (Ed.), *Constructive Drinking: Perspectives on Drink from Anthropology* (pp. 91–101). Cambridge: Cambridge University Press.

Mars, Gerald, and Yochanan Altman. (1987). "Alternative Mechanism of Distribution in a Soviet Economy." In Mary Douglas (Ed.), *Constructive Drinking: Perspectives on Drink from Anthropology* (pp. 270–279). Cambridge: Cambridge University Press.

Marshall, Alfred. (1890). *Principles of Economics.* London: MacMillan and Co.

Martin, A. Lynn. (2006). "Drinking and Alehouses in the Diary of an English Mercer's Apprentice, 1663–1674." In Mack Holt (Ed.), *Alcohol: A Social and Cultural History* (pp. 93–106). Oxford: Berg.

Mass Observation. (1943). *The Pub and the People: A Worktown Study.* London: Victor Gollancz.

Matthee, Rudolph. (2014). "Alcohol in the Islamic Middle East: Ambivalence and ambiguity." *Past and Present,* 222, 100–125.

Mattice, Sarah. (2011). "Drinking to get drunk: Pleasure, creativity, and social harmony in Greece and China." *Comparative and Continental Philosophy,* 3(2), 243–253.

Maurer, Ronald L., V. K. Kumar, Lisa Woodside, and Ronald J. Pekala. (1997). "Phenomenological experience in response to monotonous drumming and hypnotizability." *American Journal of Clinical Hypnosis,* 40(2), 130–145.

Maynard, Olivia M., Andrew L. Skinner, David M. Troy, Angela S. Attwood, and Marcus R. Munafò. (2015). "Association of alcohol consumption with perception of attractiveness in a naturalistic environment." *Alcohol and Alcoholism,* 51(2), 142–147.

McCauley, Robert N., and E. Thomas Lawson. (2002). *Bringing Ritual to Mind: Psychological Foundations of Cultural Forms.* Cambridge: Cambridge University Press.

McGovern, Patrick. (2009). *Uncorking the Past: The Quest for Wine, Beer, and Other Alcoholic Beverages.* Berkeley: University of California Press.

McGovern, Patrick. (2020). "Uncorking the Past: Alcoholic Fermentation as Humankind's First Biotechnology." In Kimberley Hockings and Robin Dunbar (Eds.), *Alcohol and Humans: A Long and Social Affair* (pp. 81–92). New York: Oxford University Press.

McKinlay, Arthur. (1951). "Attic temperance." *Quarterly Journal of Studies on Alcohol,* 12, 61–102.

McShane, Angela. (2014). "Material Culture and 'Political Drinking' in Seventeenth-Century England." In Phil Withington and Angela McShane (Eds.), *Cultures of Intoxication, Past and Present* (Vol. 222, pp. 247–276). Oxford: Oxford University Press.

Meade Eggleston, A., K. Woolaway-Bickel, and N. B. Schmidt. (2004). "Social anxiety and alcohol use: Evaluation of the moderating and mediating effects of alcohol expectancies." *Journal of Anxiety Disorders,* 18(1), 33–49.

Mehr, Samuel A., Manvir Singh, Dean Knox, Daniel M. Ketter, Daniel Pickens-Jones,

S. Atwood,...Luke Glowacki. (2019). "Universality and diversity in human song." *Science*, 366(6468), eaax0868.

Michalowski, Piotr. (1994). "The Drinking Gods: Alcohol in Early Mesopotamian Ritual and Mythology." In Lucio Milano (Ed.), *Drinking in Ancient Societies: History and Culture of Drinks in the Ancient Near East* (pp. 27–44). Padua: Sargon.

Milan, Neil F., Balint Z. Kacsoh, and Todd A. Schlenke. (2012). "Alcohol consumption as self-medication against blood-borne parasites in the fruit fly." *Current Biology*, 22(6), 488–493.

Miller, Earl, and Jonathan Cohen. (2001). "An integrative theory of prefrontal cortex function." *Annual Review of Neuroscience*, 24, 167–202.

Milton, Katharine. (2004). "Ferment in the family tree: Does a frugivorous dietary heritage influence contemporary patterns of human ethanol use?" *Integrative and Comparative Biology*, 44(4), 304–314.

Miner, Earl. (1968). *An Introduction to Japanese Court Poetry*. Palo Alto, CA: Stanford University Press.

Moeran, Brian. (2005). "Drinking Country: Flows of Exchange in a Japanese Valley." In Thomas Wilson (Ed.), *Drinking Cultures: Alcohol and Identity* (pp. 25–42). Oxford: Berg.

Mooneyham, Benjamin W., and Jonathan W. Schooler. (2013). "The costs and benefits of mind-wandering: A review." *Canadian Journal of Experimental Psychology = Revue Canadienne de Psychologie Experimentale*, 67(1), 11–18.

Morris, Steve, David Humphreys, and Dan Reynolds. (2006). "Myth, marula, and elephant: An assessment of voluntary ethanol intoxication of the African elephant (Loxodonta africana) following feeding on the fruit of the marula tree (Sclerocarya birrea)." *Physiological and Biochemical Zoology: Ecological and Evolutionary Approaches*, 79(2), 363–369.

Mountain, Mary A., and William G. Snow. (1993). "Wisconsin Card Sorting Test as a measure of frontal pathology: A review." *Clinical Neuropsychologist*, 7(1), 108–118.

Müller, Christian, and Gunter Schumann. (2011). "Drugs as instruments: A new framework for non-addictive psychoactive drug use." *Behavioral and Brain Sciences*, 34, 293–310.

Muthukrishna, Michael, Michael Doebeli, Maciej Chudek, and Joseph Henrich. (2018). "The Cultural Brain Hypothesis: How culture drives brain expansion, sociality, and life history." *PLOS Computational Biology*, 14(11), e1006504.

Nagaraja, H. S., and P. S. Jeganathan. (2003). "Effect of acute and chronic conditions of over-crowding on free choice ethanol intake in rats." *Indian Journal of Physiology and Pharmacology*, 47(3), 325–331.

Nelson, L. D., C. J. Patrick, P. Collins, A. R. Lang, and E. M. Bernat. (2011). "Alcohol impairs brain reactivity to explicit loss feedback." *Psychopharmacology*, 218(2), 419–428.

Nemeth, Z., R. Urban, E. Kuntsche, E. M. San Pedro, J. G. Roales Nieto, J. Farkas,...Z. Demetrovics. (2011). "Drinking motives among Spanish and Hungarian young adults: A cross-national study." *Alcohol*, 46(3), 261–269.

Nesse, Randolph, and Kent Berridge. (1997). "Psychoactive drug use in evolutionary perspective." *Science*, 278(5335), 63–66.

Netting, Robert. (1964). "Beer as a locus of value among the West African Kofyar." *American Anthropologist*, 66, 375–384.

Newberg, Andrew, Nancy Wintering, Donna Morgan, and Mark Waldman. (2006). "The measurement of regional cerebral blood flow during glossolalia." *Psychiatry Research,* 148, 67–71.

Nezlek, John, Constance Pilkington, and Kathryn Bilbro. (1994). "Moderation in excess: Binge drinking and social interaction among college students." *Journal of Studies on Alcohol,* 55, 342–351.

Ng Fat, Linda, Nicola Shelton, and Noriko Cable. (2018). "Investigating the growing trend of non-drinking among young people: Analysis of repeated cross-sectional surveys in England 2005–2015." *BMC Public Health,* 18(1), 1090.

Nie, Zhiguo, Paul Schweitzer, Amanda J. Roberts, Samuel G. Madamba, Scott D. Moore, and George Robert Siggins. (2004). "Ethanol augments GABAergic transmission in the central amygdala via CRF1 receptors." *Science,* 303(5663), 1512–1514.

Nietzsche, Friedrich. (1872/1967). *The Birth of Tragedy* (Walter Kaufmann, Trans.). New York: Vintage.

Nietzsche, Friedrich. (1882/1974). *The Gay Science: With a Prelude in Rhymes and an Appendix of Songs* (Walter Kaufmann, Trans.). New York: Vintage.

Nietzsche, Friedrich. (1891/1961). *Thus Spoke Zarathustra* (R. J. Hollingdale, Trans.). New York: Penguin.

Norenzayan, Ara, Azim Shariff, William Gervais, Aiyana Willard, Rita McNamara, Edward Slingerland, and Joseph Henrich. (2016). "The cultural evolution of prosocial religions." *Behavioral and Brain Sciences,* 39, e1 (19 pages).

Norris, J., and K. L. Kerr. (1993). "Alcohol and violent pornography: Responses to permissive and nonpermissive cues." *Journal of Studies on Alcohol, Supplement,* 11, 118–127.

Nugent, Paul. (2014). "Modernity, Tradition, and Intoxication: Comparative Lessons from South Africa and West Africa." In Phil Withington and Angela McShane (Eds.), *Cultures of Intoxication, Past and Present* (Vol. 222, pp. 126–145). Oxford: Oxford University Press.

O'Brien, Sara Ashley. (2016, February 26). "Zenefits Lays Off 250 Employees." *CNN.*

O'Brien, Sara Ashley. (2018, October 31). "WeWork to Limit Free Beer All-Day Perk to Four Glasses." *CNN.*

O'Connor, Anahad. (2020, July 10). "Should We Be Drinking Less? Scientists Helping to Update the Latest Edition of the Dietary Guidelines for Americans Are Taking a Harder Stance on Alcohol." *New York Times.*

Olive, M. Foster, Heather N. Koenig, Michelle A. Nannini, and Clyde W. Hodge. (2001). "Stimulation of endorphin neurotransmission in the nucleus accumbens by ethanol, cocaine, and amphetamine." *Journal of Neuroscience,* 21(23), RC184.

Olsen, Richard W., Harry J. Hanchar, Pratap Meera, and Martin Wallner. (2007). "GABAA receptor subtypes: The 'one glass of wine' receptors." *Alcohol,* 41(3), 201–209.

"On the Road Again: Companies Are Spending More on Sending Their Staff Out to Win Deals." (2015, November 21). *The Economist.*

Orehek E., L. Human, M. A. Sayette, J. D. Dimoff, R. P. Winograd, and K. J. Sher. (2020). "Self-expression while drinking alcohol: Alcohol influences personality expression during first impressions." *Personality and Social Psychology Bulletin,* 46(1), 109–123.

Oroszi, Gabor, and David Goldman. (2004). "Alcoholism: Genes and mechanisms." *Pharmacogenomics*, 5(8), 1037–1048.

Osborne, Robin. (2014). "Intoxication and Sociality: The Symposium in the Ancient Greek World." In Phil Withington and Angela McShane (Eds.), *Cultures of Intoxication, Past and Present* (Vol. 222, pp. 34–60). Oxford: Oxford University Press.

Pahnke, Walter. (1963). *Drugs and Mysticism: An Analysis of the Relationship Between Psychedelic Drugs and the Mystical Consciousness.* (Ph.D. dissertation.) Cambridge, MA: Harvard University Press.

Park, Seung Kyu, Choon-Sik Park, Hyo-Suk Lee, Kyong Soo Park, Byung Lae Park, Hyun Sub Cheong, and Hyoung Doo Shin. (2014). "Functional polymorphism in aldehyde dehydrogenase-2 gene associated with risk of tuberculosis." *BMC Medical Genetics,* 15(1), 40.

Patrick, Clarence H. (1952). *Alcohol, Culture and Society.* Durham, NC: Duke University Press.

Peele, Stanton, and Archie Brodsky. (2000). "Exploring psychological benefits associated with moderate alcohol use: A necessary corrective to assessments of drinking outcomes?" *Drug and Alcohol Dependence*, 60(3), 221–247.

Peng, G. S., Y. C. Chen, M. F. Wang, C. L. Lai, and S. J. Yin. (2014). "ALDH2*2 but not ADH1B*2 is a causative variant gene allele for Asian alcohol flushing after a low-dose challenge: Correlation of the pharmacokinetic and pharmacodynamic findings." *Pharmacogenetics and Genomics*, 24(12), 607–617.

Peng, Yi, Hong Shi, Xue-bin Qi, Chun-jie Xiao, Hua Zhong, Run-lin Z. Ma, and Bing Su. (2010). "The ADH1B Arg47His polymorphism in East Asian populations and expansion of rice domestication in history." *BMC Evolutionary Biology*, 10(1), 15.

Pinker, Steven. (1997). *How the Mind Works.* New York: W. W. Norton & Company.

Platt, B. S. (1955). "Some traditional alcoholic beverages and their importance in indigenous African communities." *Proceedings of the Nutrition Society*, 14, 115–124.

Polimanti, Renato, and Joel Gelernter. (2017). ADH1B: "From alcoholism, natural selection, and cancer to the human phenome." *American Journal of Medical Genetics*, 177(2), 113–125.

Polito, V., and R. J. Stevenson. (2019). "A systematic study of microdosing psychedelics." *PLOS ONE*, 14(2), e0211023.

Pollan, Michael. (2001). *The Botany of Desire: A Plant's-Eye View of the World.* New York: Random House.

Pollan, Michael. (2018). *How to Change Your Mind.* New York: Penguin.

Poo, Mu-chou. (1999). "The use and abuse of wine in ancient China." *Journal of the Economic and Social History of the Orient*, 42(2), 123–151.

Porter, Stephen, Leanne ten Brinke, Alysha Baker, and Brendan Wallace. (2011). "Would I lie to you? 'Leakage' in deceptive facial expressions relates to psychopathy and emotional intelligence." *Personality and Individual Differences*, 51, 133–137.

Powers, Madelon. (2006). "The Lore of the Brotherhood: Continuity and Change in the Urban American Saloon Cultures, 1870–1920." In Mack Holt (Ed.), *Alcohol: A Social and Cultural History* (pp. 145–160). Oxford: Berg.

Price, N. (2002). *The Viking Way: Religion and War in Late Iron Age Scandinavia.* Uppsala: University of Uppsala Press.

Prochazkova, L., D. P. Lippelt, L. S. Colzato, M. Kuchar, Z. Sjoerds, and B. Hommel. (2018). "Exploring the effect of microdosing psychedelics on creativity in an open-label natural setting." *Psychopharmacology,* 235(12), 3401–3413.

Radcliffe-Brown, A. R. (1922/1964). *The Andaman Islanders.* New York: Free Press.

Rand, David. (2019, May 17). "Intuition, deliberation, and cooperation: Further meta-analytic evidence from 91 experiments on pure cooperation." *Social Science Research Network.* Available at SSRN: https://ssrn.com/abstract=3390018.

Rand, David, Joshua Greene, and Martin Nowak. (2012). "Spontaneous giving and calculated greed." *Nature,* 489(7416), 427–430.

Rappaport, Roy A. (1999). *Ritual and Religion in the Making of Humanity.* Cambridge: Cambridge University Press.

Raz, Gil. (2013). "Imbibing the universe: Methods of ingesting the five sprouts." *Asian Medicine,* 7, 76–111.

Reddish, Paul, Joseph Bulbulia, and Ronald Fischer. (2013). "Does synchrony promote generalized prosociality?" *Religion, Brain and Behavior,* 4(1), 3–19.

Reinhart, Katrinka. (2015). "Religion, violence, and emotion: Modes of religiosity in the Neolithic and Bronze Age of Northern China." *Journal of World Prehistory,* 28(2), 113–177.

Richerson, Peter J., and Robert Boyd. (2005). *Not by Genes Alone: How Culture Transformed Human Evolution.* Chicago: University of Chicago Press.

Riemer, Abigail R., Michelle Haikalis, Molly R. Franz, Michael D. Dodd, David DiLillo, and Sarah J. Gervais. (2018). "Beauty is in the eye of the beer holder: An initial investigation of the effects of alcohol, attractiveness, warmth, and competence on the objectifying gaze in men." *Sex Roles,* 79(7), 449–463.

Rogers, Adam. (2014). *Proof: The Science of Booze.* Boston: Houghton Mifflin.

Rosinger, Asher, and Hilary Bathancourt. (2020). "*Chicha* as water: Traditional fermented beer consumption among forager-horticulturalists in the Bolivian Amazon." In Kimberley Hockings and Robin Dunbar (Eds.), *Alcohol and Humans: A Long and Social Affair* (pp. 147–162). New York: Oxford University Press.

Roth, Marty. (2005). *Drunk the Night Before: An Anatomy of Intoxication.* Minneapolis: University of Minnesota Press.

Rucker, James J. H., Jonathan Iliff, and David J. Nutt. (2018). "Psychiatry and the psychedelic drugs. Past, present and future." *Neuropharmacology,* 142, 200–218.

Rudgley, Richard. (1993). *Alchemy of Culture: Intoxicants in Society.* London: British Museum Press.

Samorini, Giorgio. (2002). *Animals and Psychedelics: The Natural World and the Instinct to Alter Consciousness.* Rochester, VT: Park Street Press.

Sanchez, F., M. Melcon, C. Korine, and B. Pinshow. (2010). "Ethanol ingestion affects flight performance and echolocation in Egyptian fruit bats." *Behavioural Processes,* 84(2), 555–558.

Sayette, Michael (1999). "Does drinking reduce stress?" *Alcohol Research and Health,* 23(4), 250–255.

Sayette, Michael, K. G. Creswell, J. D. Dimoff, C. E. Fairbairn, J. F. Cohn, B. W. Heckman,...R. L. Moreland. (2012). "Alcohol and group formation: A multimodal investigation of the effects of alcohol on emotion and social bonding." *Psychological Science*, 23(8), 869–878.

Sayette, Michael A., Erik D. Reichle, and Jonathan W. Schooler. (2009). "Lost in the sauce: The effects of alcohol on mind wandering." *Psychological Science*, 20(6), 747–752.

Schaberg, David. (2001). *A Patterned Past: Form and Thought in Early Chinese Historiography*. Cambridge, MA: Harvard University Press.

Schivelbusch, Wolfgang. (1993). *Tastes of Paradise: A Social History of Spices, Stimulants, and Intoxicants* (David Jacobson, Trans.). New York: Vintage Books.

Schmidt, K. L., Z. Ambadar, J. F. Cohn, and L. I. Reed. (2006). "Movement differences between deliberate and spontaneous facial expressions: Zygomaticus major action in smiling." *Journal of Nonverbal Behavior*, 30(1), 37–52.

Schuckit, Marc A. (2014). "A brief history of research on the genetics of alcohol and other drug use disorders." *Journal of Studies on Alcohol and Drugs*, 75(Supplement 17), 59–67.

Sharon, Douglas. (1972). "The San Pedro Cactus in Peruvian Folk Healing." In Peter Furst (Ed.), *Flesh of the Gods: The Ritual Use of Hallucinogens* (pp. 114–135). New York: Praeger.

Shaver, J. H., and R. Sosis. (2014). "How does male ritual behavior vary across the lifespan? An examination of Fijian kava ceremonies." *Human Nature*, 25(1), 136–160.

Sher, Kenneth, and Mark Wood. (2005). "Subjective Effects of Alcohol II: Individual Differences." In Mitch Earleywine (Ed.), *Mind-Altering Drugs: The Science of Subjective Experience* (pp. 135–153). New York: Oxford University Press.

Sher, Kenneth, Mark Wood, Alison Richardson, and Kristina Jackson. (2005). "Subjective Effects of Alcohol I: Effects of the Drink and Drinking Context." In Mitch Earleywine (Ed.), *Mind-Altering Drugs: The Science of Subjective Experience* (pp. 86–134). New York: Oxford University Press.

Sherratt, Andrew. (2005). "Alcohol and Its Alternatives: Symbol and Substance in Pre-Industrial Cultures." In Jordan Goodman, Andrew Sherratt, and Paul E. Lovejoy (Eds.), *Consuming Habits: Drugs in History and Anthropology* (pp. 11–46). New York: Routledge.

Shohat-Ophir, G., K. R. Kaun, R. Azanchi, H. Mohammed, and U. Heberlein. (2012). "Sexual deprivation increases ethanol intake in Drosophila." *Science*, 335(6074), 1351–1355.

Shonle, Ruth. (1925). "Peyote: The giver of visions." *American Anthropologist*, 27, 53–75.

Sicard, Delphine, and Jean-Luc Legras. (2011). "Bread, beer and wine: Yeast domestication in the Saccharomyces sensu stricto complex." *Comptes Rendus Biologies*, 334(3), 229–236.

Siegel, Jenifer, and Molly Crockett. (2013). "How serotonin shapes moral judgment and behavior." *Annals of the New York Academy of Sciences*, 1299(1), 42–51.

Siegel, Ronald. (2005). *Intoxication: The Universal Drive for Mind-Altering Substances*. Rochester, VT: Park Street Press.

Silk, Joan. (2002). "Grunts, Girneys, and Good Intentions: The Origins of Strategic Commitment in Nonhuman Primates." In Randolph M. Nesse (Ed.),

Evolution and the Capacity for Commitment (pp. 138–157). New York: Russell Sage Foundation.

Skyrms, Brian. (2004). *The Stag Hunt and the Evolution of Social Structure.* Cambridge: Cambridge University Press.

Slingerland, Edward. (2008a). "The problem of moral spontaneity in the Guodian corpus. *Dao: A Journal of Comparative Philosophy,* 7(3), 237–256.

Slingerland, Edward. (2008b). *What Science Offers the Humanities: Integrating Body and Culture.* New York: Cambridge University Press.

Slingerland, Edward. (2014). *Trying Not to Try: Ancient China, Modern Science and the Power of Spontaneity.* New York: Crown Publishing.

Slingerland, Edward, and Mark Collard. (2012). "Creating Consilience: Toward a Second Wave." In Edward Slingerland and Mark Collard (Eds.), *Creating Consilience: Integrating the Sciences and the Humanities* (pp. 3–40). New York: Oxford University Press.

Smail, Daniel Lord. (2007). *On Deep History and the Brain.* Berkeley: University of California Press.

Smith, Huston. (1964). "Do drugs have religious import?" *Journal of Philosophy,* 61(18), 517–530.

Sommer, Jeffrey D. (1999). "The Shanidar IV 'flower burial': A re-evaluation of Neanderthal burial ritual." *Cambridge Archaeological Journal,* 9(1), 127–129.

Sophocles. (1949). *Oedipus Rex* (Dudley Fitts and Robert Fitzgerald, Trans.). New York: Harcourt Brace.

Sowell, E. R., D. A. Trauner, A. Gamst, and T. L. Jernigan. (2002). "Development of cortical and subcortical brain structures in childhood and adolescence: A structural MRI study." *Developmental Medicine & Child Neurology,* 44(1), 4–16.

Sowles, Kara. (2014, October 28). "Alcohol and inclusivity: Planning tech events with non-alcoholic options." *Model View Culture.*

Sparks, Adam, Tyler Burleigh, and Pat Barclay. (2016). "We can see inside: Accurate prediction of Prisoner's Dilemma decisions in announced games following a face-to-face interaction." *Evolution and Human Behavior,* 37(3), 210–216.

Spinka, Marek, Ruth C. Newberry, and Marc Bekoff. (2001). "Mammalian play: Training for the unexpected. *Quarterly Review of Biology,* 76(2), 141–168.

St. John, Graham (Ed.). (2004). *Rave Culture and Religion.* London: Routledge.

Staal, Frits. (2001). "How a psychoactive substance becomes a ritual: The case of soma." *Social Research,* 68(3), 745–778.

Steele, Claude M., and Robert A. Josephs. (1990). "Alcohol myopia: Its prized and dangerous effects." *American Psychologist,* 45(8), 921–933.

Steinkraus, Keith H. (1994). "Nutritional significance of fermented foods." *Food Research International,* 27(3), 259–267.

Sterckx, Roel. (2006). "Sages, cooks, and flavours in Warring States and Han China." *Monumenta Serica,* 54, 1-47.

Studerus, Erich, Alex Gamma, and Franz X. Vollenweider. (2010). "Psychometric evaluation of the altered states of consciousness rating scale (OAV)." *PLOS ONE,* 5(8), e12412.

Sullivan, Roger J., Edward H. Hagen, and Peter Hammerstein. (2008). "Revealing the paradox of drug reward in human evolution." *Proceedings of the Royal Society B: Biological Sciences,* 275(1640), 1231–1241.

Szaif, Jan. (2019). "Drunkenness as a communal practice: Platonic and peripatetic perspectives." *Frontiers of Philosophy in China,* 14(1), 94–110.

Talin, P., and E. Sanabria. (2017). "Ayahuasca's entwined efficacy: An ethnographic study of ritual healing from 'addiction.'" *International Journal of Drug Policy,* 44, 23–30.

Tarr, B., J. Launay, and R. I. Dunbar. (2016). "Silent disco: Dancing in synchrony leads to elevated pain thresholds and social closeness." *Evolution and Human Behavior,* 37(5), 343–349.

Taylor, B., H. M. Irving, F. Kanteres, R. Room, G. Borges, C. Cherpitel,...J. Rehm. (2010). "The more you drink, the harder you fall: A systematic review and meta-analysis of how acute alcohol consumption and injury or collision risk increase together." *Drug and Alcohol Dependence,* 110(1-2), 108–116.

Taylor, Jenny, Naomi Fulop, and John Green. (1999). "Drink, illicit drugs and unsafe sex in women." *Addiction,* 94(8), 1209–1218.

ten Brinke, Leanne, Stephen Porter, and Alysha Baker. (2012). "Darwin the detective: Observable facial muscle contractions reveal emotional high-stakes lies." *Evolution and Human Behavior,* 33(4), 411–416.

ten Brinke, Leanne, K. D. Vohs, and D. R. Carney. (2016). "Can ordinary people detect deception after all?" *Trends in Cognitive Sciences,* 20(8), 579–588.

Testa, M., C. A. Crane, B. M. Quigley, A. Levitt, and K. E. Leonard. (2014). "Effects of administered alcohol on intimate partner interactions in a conflict resolution paradigm." *Journal of Studies on Alcohol and Drugs,* 75(2), 249–258.

Thompson-Schill, Sharon, Michael Ramscar, and Evangelia Chrysikou. (2009). "Cognition without control: When a little frontal lobe goes a long way." *Current Directions in Psychological Science,* 18(5), 259–263.

Tlusty, B. Ann. (2001). *Bacchus and Civic Order: The Culture of Drink in Early Modern Germany.* Charlottesville: University of Virginia Press.

"To Your Good Stealth: A Beery Club of Euro-Spies That Never Spilt Secrets." (2020, May 30). *The Economist.*

Todorov, Alexander, Manish Pakrashi, and Nikolaas N. Oosterhof. (2009). "Evaluating faces on trustworthiness after minimal time exposure." *Social Cognition,* 27(6), 813–833.

Tognetti, Arnaud, Claire Berticat, Michel Raymond, and Charlotte Faurie. (2013). "Is cooperativeness readable in static facial features? An inter-cultural approach." *Evolution and Human Behavior,* 34(6), 427–432.

Tooby, John, and Leda Cosmides. (2008). "The Evolutionary Psychology of the Emotions and Their Relationship to Internal Regulatory Variables." In Michael Lewis, Jeannette M. Haviland-Jones, and Lisa Feldman Barrett (Eds.), *Handbook of Emotion* (Third Edition, pp. 114–137). New York: Guilford Press.

Toren, Christina. (1988). "Making the present, revealing the past: The mutability and continuity of tradition as process." *Man,* 23, 696.

Tracy, Jessica, and Richard Robbins. (2008). "The automaticity of emotion recognition." *Emotion,* 8(1), 81–95.

Tramacchi, Des. (2004). "Entheogenic Dance Ecstasis: Cross-Cultural Contexts." In Graham St. John (Ed.), *Rave Culture and Religion* (pp. 125–144). London: Routledge.

Turner, Fred. (2009). "Burning Man at Google: A cultural infrastructure for new media production." *New Media and Society*, 11(1-2), 145–166.

Vaitl, Dieter, John Gruzelier, Graham A. Jamieson, Dietrich Lehmann, Ulrich Ott, Gebhard Sammer,...Thomas Weiss. (2005). "Psychobiology of altered states of consciousness." *Psychological Bulletin*, 131(1), 98–127.

Vallee, Bert L. (1998). "Alcohol in the Western world." *Scientific American*, 278(6), 80–85.

Van den Abbeele, J., I. S. Penton-Voak, A. S. Attwood, I. D. Stephen, and M. R. Munafò. (2015). "Increased facial attractiveness following moderate, but not high, alcohol consumption." *Alcohol and Alcoholism*, 50(3), 296–301.

van't Wout, M., and A. G. Sanfey. (2008). "Friend or foe: The effect of implicit trustworthiness judgments in social decision-making." *Cognition*, 108(3), 796–803.

Veit, Lena, and Andreas Nieder. (2013). "Abstract rule neurons in the endbrain support intelligent behaviour in corvid songbirds." *Nature Communications*, 4(1), 2878.

Wadley, Greg. (2016). "How psychoactive drugs shape human culture: A multidisciplinary perspective." *Brain Research Bulletin*, 126(Part 1), 138–151.

Wadley, Greg, and Brian Hayden. (2015). "Pharmacological influences on the Neolithic transition." *Journal of Ethnobiology*, 35(3), 566–584.

Waley, Arthur. (1996). *The Book of Songs: The Ancient Chinese Classic of Poetry*. New York: Grove Press.

Walton, Stuart. (2001). *Out of It: A Cultural History of Intoxication*. London: Penguin.

Wang, Jiajing, Li Liu, Terry Ball, Linjie Yu, Yuanqing Li, and Fulai Xing. (2016). "Revealing a 5,000-year-old beer recipe in China." *Proceedings of the National Academy of Sciences*, 113(23), 6444.

Warrington, Ruby. (2018). *Sober Curious: The Blissful Sleep, Greater Focus, Limitless Presence, and Deep Connection Awaiting Us All on the Other Side of Alcohol*. New York: HarperOne.

Wasson, R. Gordon. (1971). "The Soma of the Rig Veda: What was it?" *Journal of the American Oriental Society*, 91(2), 169–187.

Watson, Burton. (1968). *The Complete Works of Chuang Tzu*. New York: Columbia University Press.

Watson, P. L., O. Luanratana, and W. J. Griffin. (1983). "The ethnopharmacology of pituri." *Journal of Ethnopharmacology*, 8(3), 303–311.

Weil, Andrew. (1972). *The Natural Mind: A New Way of Looking at Drugs and the Higher Consciousness*. Boston: Houghton Mifflin.

Weismantel, Mary. (1988). *Food, Gender, and Poverty in the Ecuadorian Andes*. Philadelphia: University of Pennsylvania Press.

Wettlaufer, Ashley, K. Vallance, C. Chow, T. Stockwell, N. Giesbrecht, N. April,...K. Thompson. (2019). *Strategies to Reduce Alcohol-Related Harms and Costs in Canada: A Review of Federal Policies*. Victoria, BC: Canadian Institute for Substance Use Research, University of Victoria.

Wheal, Jamie, and Steven Kotler. (2017). *Stealing Fire: How Silicon Valley, the Navy SEALs, and Maverick Scientists Are Revolutionizing the Way We Live and Work*. New York: Dey Street Books.

Whitehouse, Harvey. (2004). *Modes of Religiosity: A Cognitive Theory of Religious Transmission*. Walnut Creek, CA; Toronto, ON: AltaMira Press.

Wiessner, P. W. (2014). "Embers of society: Firelight talk among the Ju/'hoansi Bushmen." *Proceedings of the National Academy of Sciences,* 111(39), 14027–14035.

Williams, Alex. (2019, June 15). "The New Sobriety." *The New York Times.*

Williams, Nicholas Morrow. (2013). "The Morality of Drunkenness in Chinese Literature of the Third Century CE." In Isaac Yue and Siufu Tang (Eds.), *Scribes of Gastronomy* (pp. 27–43). Hong Kong: Hong Kong University Press.

Willis, Janine, and Alexander Todorov. (2006). "First impressions: Making up your mind after a 100-ms exposure to a face." *Psychological Science,* 17(7), 592–598.

Willoughby, Laura, Jussi Tolvi, and Dru Jaeger. (2019). *How to Be a Mindful Drinker: Cut Down, Stop for a Bit, or Quit.* London: DK Publishing.

Wilson, Bundy, N. J. Mackintosh, and R. A. Boakes. (1985). "Transfer of relational rules in matching and oddity learning by pigeons and corvids." *Quarterly Journal of Experimental Psychology Section B,* 37(4b), 313–332.

Wilson, Carla. (2019, January 7). "B.C. Scientist Heads Survey into Secret Lives of Pacific Salmon." *Vancouver Sun.*

Wilson, David Sloan. (2007). *Evolution for Everyone: How Darwin's Theory Can Change the Way We Think About Our Lives.* New York: Delacorte Press.

Wilson, Thomas. (2005). "Drinking Cultures: Sites and Practices in the Production and Expression of Identity." In Thomas Wilson (Ed.), *Drinking Cultures: Alcohol and Identity* (pp. 1–25). Oxford: Berg.

Winkelman, Michael. (2002). "Shamanism as neurotheology and evolutionary psychology." *American Behavioral Scientist,* 45, 1875–1887.

Wise, R. A. (2000). "Addiction becomes a brain disease." *Neuron,* 26(1), 27–33.

Wood, R. M., J. K. Rilling, A. G. Sanfey, Z. Bhagwagar, and R. D. Rogers. (2006). "Effects of tryptophan depletion on the performance of an iterated Prisoner's Dilemma game in healthy adults." *Neuropsychopharmacology,* 31(5), 1075–1084.

World Health Organization. (2018). "Global status report on alcohol and health 2018."

Wrangham, Richard. (2009). *Catching Fire: How Cooking Made Us Human.* New York: Basic Books.

Yan, Ge. (2019, November 30). "How to Survive as a Woman at a Chinese Banquet." *The New York Times.*

Yanai, Itai, and Martin Lercher. (2016). *The Society of Genes.* Cambridge, MA: Harvard University Press.

Yong, Ed. (2018, June 21). "A Landmark Study on the Origins of Alcoholism." *The Atlantic.*

Young, Chelsie M., Angelo M. DiBello, Zachary K. Traylor, Michael J. Zvolensky, and Clayton Neighbors. (2015). "A longitudinal examination of the associations between shyness, drinking motives, alcohol use, and alcohol-related problems." *Alcoholism: Clinical and Experimental Research,* 39(9), 1749–1755.

Zabelina, Darya L., and Michael D. Robinson. (2010). "Child's play: Facilitating the originality of creative output by a priming manipulation." *Psychology of Aesthetics, Creativity, and the Arts,* 4(1), 57–65.

NOTES

INTRODUCTION

1 The author Michael Pollan refers to these as "transparent" drugs, "whose effects on consciousness are too subtle to interfere with one's ability to get through the day and fulfill one's obligations. Drugs such as coffee, tea, and tobacco in our culture, or coca or khat leaves in others, leave the user's space-time coordinates untouched" (Pollan 2018: 142). Stephen Braun similarly distinguishes between "normalizing" drugs, like caffeine or nicotine, and "intoxicating" ones (Braun 1996: 164).

2 Although they are working with a much more expansive sense of "altered states" than the one I will be using in this book (encompassing porn, gambling, and other forms of immersive entertainment, as well as stimulants such as tobacco or caffeine), the authors of *Stealing Fire* (Wheal and Kotler 2017) estimate that, worldwide, humans spend approximately 4 trillion dollars per year (2016 U.S. dollars) solely in the pursuit of "getting out of [their] heads."

3 McGovern 2009.

4 Excellent accounts of the history of intoxication can be found in Curry 2017; Forsyth 2017; Gately 2008; Guerra-Doce 2014; McGovern 2009, 2020; Sherratt 2005; Vallee 1998; Walton 2001.

5 The classic "beer before bread" argument can be found in Braidwood et al. 1953; cf. Katz and Voight 1986 and Dietler 2006. We'll return to this theory in Chapter Three.

6 Translation from http://www2.latech.edu/~bmagee/103/gilgamesh.htm; also see A. George 2003: 12-15.

7 This theory that soma was made from *Amanita muscaria*, a hallucinogenic mushroom, was most famously and enthusiastically advanced by an amateur mycologist and philologist named Gordon Wasson (Wasson 1971), who managed to convince many scholars of early Vedic culture. Wendy Doniger (Doniger O'Flaherty 1968) provides a comprehensible review of various theories of the identity of *soma;* also see the discussion in Staal 2001.

8 *Rig Veda* 10.119, translation by Wendy Doniger as modified by Fritz Staal (Staal 2001: 751-752).

9 The most fun and readable of the ones mentioned above are Forsyth 2017; Gately 2008; Walton 2001. Gately 2008 is probably the most comprehensive resource for the history of alcohol use, and was very useful in the early stages of this book

project. While Gately mentions in passing some potential individual- and social-level functions of alcohol, there is no attempt to provide a rigorous psychological, neuro-biological, genetic, or cultural evolutionary explanation for the phenomenon. Forsyth 2017, which came out in the early stages of writing this book, can be seen as a shorter and more deliberately humorous version of Gately's comprehensive work. Forsyth does begin with the *why* question about our drive for intoxication. He, however, quickly and rather uncritically embraces the "drunken monkey" hypothesis of Robert Dudley ("We humans are champion booze hounds and the Drunken Monkey Hypothesis explains why" (2017: 15)), which fits uneasily with the history of drunkenness he goes on to relate, as I will explain in Chapter One.

As we'll discuss more, in the academic world there have been some anthropologists and archaeologists who have advanced functional accounts of intoxicant use, most notably Dietler 2006; O. Dietrich et al. 2012; Robin Dunbar 2014; Robin Dunbar et al. 2016; Guerra-Doce 2014; E. Hagen and Tushingham 2019; D. Heath 2000; Jennings and Bowser 2009; McGovern 2009; Wadley and Hayden 2015. Perhaps the most important collection of essays on this topic (Hockings and Dunbar 2020) was published during the final stages of the writing of this book; Heath 2000: Ch. 6 ("The Heart Has Its Reasons: Why Do People Drink?") is also highly recommended. We will also explore the evolutionary hypotheses of biologists such as Robert Dudley and Matthew Carrigan; for a helpful brief survey of such theories, see M. Carrigan 2020: 24-25 or McGovern 2020: 86-87. More typically, however, chemical intoxicants are either ignored in anthropological circles or treated as cultural "signifiers" detached from any underlying effects on human psychology. See, for instance, MacAndrew and Edgerton's comment that it is "likely" that alcohol intake itself has no cognitive disinhibiting function, merely some motor effects that then serve as a visible social symbol of whatever intoxication is constructed as meaning in that culture (MacAndrew and Edgerton 1969). Dietler 2006 provides an overview of the history of anthropological approaches to alcohol, although he sees the most recent, cultural constructivist phase as an advance over the functionalism of the 1970s and '80s. Finally, for a landmark collection of anthropological essays representing the more exclusively cultural views on alcohol consumption that have dominated anthropology from the 1980s through the present day, see Douglas 1987.

10   As should become clear in the pages that follow, I will be working within the theoretical framework of what has variously been called gene-culture coevolution (Richerson and Boyd 2005) or dual inheritance theory (Henrich and McElreath 2007), which sees human cognition and behavior as driven by two different modes of inheritance, genetic and cultural. I will therefore generally be using the term "evolutionary" to refer to both genetic and cultural evolution, although I will be more specific when necessary. The worries expressed by scholars, such as Michael Dietler, about "overly facile direct invocation of genetic or evolutionary explanations" (2020: 125) when it comes to cultural attitudes toward alcohol incorrectly conflate "evolutionary" with "genetic." While some theorists of alcohol use may be working from a more narrow, and outmoded, "genes only" perspective, the gene-culture coevolutionary framework is arguably the standard model in evolutionary approaches to human behavior these days (see, e.g., Henrich 2015; Norenzayan et al. 2016; Slingerland and Collard 2012).

11   See, e.g., Gerbault et al. 2011.

12   Griffith Edwards 2000: 56, who attributes its success to its relatively moderate intoxicating effects and its ability to be shaped and regulated by cultural norms, unlike more powerful intoxicating drugs (56-57). Also see Sher and Wood 2005 on alcohol's uniquely predictable, dose-responsive effects (as opposed to, for instance, cannabis; Kuhn and Swartzwelder 1998: 181) and Mäkelä 1983 on its easy integration into other cultural practices.

CHAPTER ONE: WHY DO WE GET DRUNK?

1   Dietler 2020: 115.
2   For overviews of the history of human drinking, see Forsyth 2017; Gately 2008; McGovern 2009.
3   Vénus à la corne de Laussel from Collection Musée d'Aquitaine, see discussion in McGovern 2009: 16-17.
4   McGovern et al. 2004; McGovern 2020. There is also evidence of beer-making in early China dating back to around 5,000 years ago. This mixture of millet, barley, and tubers would be unlikely to win any prizes in a contemporary beer festival, but these early brewers appear to have been tinkering with recipes before the widespread establishment of settled agriculture in the region (Wang et al. 2016).
5   Gately 2008: 3.
6   Barnard et al. 2011.
7   Dineley 2004.
8   Kirkby 2006: 212.
9   Hagen and Tushingham 2019; Sherratt 2005.
10   Rucker, Iliff, and Nutt 2018.
11   See Carod-Artal 2015; Furst 1972 on "mushroom stones"; Sharon 1972: 115-116 on "San Pedro" cactus on ceramic vessel from the Chavín culture (1200-600 BCE).
12   The relevant toxins, collectively referred to as "bufotoxins," are produced by toads of the genus *Bufo*; see Carod-Artal 2015.
13   Joseph Henrich (personal communication) has speculated that, because there is some evidence that alcohol might worsen the effects of seafood poisoning caused by ciguatera, a toxic microorganism that affects reef fish, cultures where ciguatera is a problem may have turned instead to kava. The distribution of kava-dominated cultures in the Pacific does seem to overlap with ciguatera prevalence.
14   Lebot, Lindstrom, and Merlin 1992: 13.
15   Long et al. 2016.
16   Hagen and Tushingham 2019.
17   See Sherratt 2005: 26-27. Cannabis was probably used in conjunction with opium throughout this region.
18   Carmody et al. 2018.
19   The strain of tobacco grown by most native cultures, *Nicotiana rustica*, is much stronger than today's commercial variety, *Nicotiana attenuata*; common hallucinogens traditionally smoked along with tobacco included datura and brugmansia (Fuller 2000: 35; Carod-Artal 2015; Schultes 1972: 46-47).
20   Dineley 2004.
21   Guerra-Doce 2014.

22   E.g., Gately 2008; Forsyth 2017.

23   Weil 1972: 14.

24   Sherratt 2005: 33. As Sherratt notes, in the central zones of the globe, where agriculture and large-scale civilization got its start, alcohol produced from grain or fruit is generally the drug of choice. More northern peoples tend to use narcotics like opium, cannabis, or tobacco, while southern peoples have developed a penchant for stimulants like cocaine, *qat,* coffee, or tea. Everywhere people are also getting high on various hallucinogens made from vines, cacti, or mushrooms (Sherratt 2005: 32).

25   There is, it should be noted, a longstanding attempt in anthropological circles to explain alcohol use in functional terms, most of which have tended to focus on anxiety or stress relief. Patrick 1952: 45-47 provides a nice summary of classic anthropological theories from the 1920s to the 1940s. We will discuss the stress-relief theory, as well as more recent anthropological attempts to explain our taste for alcohol, below.

26   R. Siegel 2005: 54.

27   From 1814, quoted in Blocker 2006: 228.

28   Nesse and Berridge 1997: 63-64.

29   As Pinker observes, "People watch pornography when they could be seeking a mate, forgo food to buy heroin, sell their blood to buy movie tickets (in India), postpone childbearing to climb the corporate ladder, and eat themselves into an early grave. Human vice is proof that biological adaptation is, speaking literally, a thing of the past. Our minds are adapted to the small foraging bands in which our family spent ninety-nine percent of its existence, not the topsy-turvey contingencies we have created since the agricultural and industrial revolutions" (Pinker 1997: 207). For other expressions of the hijack hypothesis, see Hyman 2005 and Wise 2000.

30   Heberlein et al. 2004.

31   Devineni and Heberlein 2009.

32   Shohat-Ophir et al. 2012.

33   That said, there are clearly adaptive forces at play in the attraction of fruit flies to alcohol and their ability to process it. Pursuing ethanol leads them to overripe fruit and therefore their primary food source and, as we will discuss below, they sometimes use their enhanced ability to process alcohol against predators, such as parasitic wasps.

34   Dudley 2014, 2020.

35   He adds, "Some humans are, in effect, abused by alcohol as it activates ancient neural pathways that were once nutritionally useful but that now falsely signal reward following excessive consumption." Dudley 2014: xii-xiii.

36   Steinkraus 1994; Battcock and Azam-Ali 1998.

37   https://www.economist.com/middle-east-and-africa/2018/02/08/what-is-cheaper -than-beer-and-gives-you-energy.

38   Fermenting maize into beer almost doubles levels of riboflavin and nicotinic acid, and triples or quadruples the level of B vitamins; turning wheat into beer produces essential amino acids, enhances B-vitamin levels, and provides substances that improve the absorption of essential minerals. Platt 1955; Steinkraus 1994; Katz and Voight 1986.

39   Curry 2017.

40 See studies cited in Chrzan 2013: 53-55.

41 See studies cited by Dietler 2020: 118.

42 Milan, Kacsoh, and Schlenke 2012.

43 Rosinger and Bathancourt 2020: 147. Also see Vallee 1998; Arthur 2014.

44 For instance, Sullivan, Hagen, and Hammerstein 2008 argue that plant neurotoxins can serve as anthelmintic drugs, which may have provided significant adaptive advantage for our ancestors, who lived shorter lives and carried heavy parasite loads without the benefits of modern medicine.

45 See, for instance, the comments of Dudley's UC Berkeley colleague Katharine Milton (Milton 2004), who also points out that non-fruit-eating mammals such as mice and rats also show patterns of ethanol consumption similar to humans. As Dudley notes in his defense (2020: 10), however, some more recent work (Peris et al. 2017) does suggest that fermentation in fruit can boost its aroma and make it more appealing to mammals and birds.

46 The best introduction to the power of cultural evolution, and our species' reliance upon it, is Henrich 2015. Also see the discussion of the importance of accumulated culture for human beings in Chapter Two.

47 Dietler 2006, in response to Joffe 1998, notes that surveys cross-culturally show that people frequently drink both alcohol and water, or mix alcohol with water. The Greeks famously diluted their wine with water.

48 For one prominent theory on what other adaptive forces are at work, see Norenzayan et al. 2016 and accompanying commentaries.

49 As Iain Gately notes, when it came to early European explorations of the world, "Wine was a significant part of the cost of fitting out an expedition. Magellan spent more on sherry than on armaments; indeed, his wine rations cost nearly twice as much as his flagship, the *San Antonio*" (Gately 2008: 95).

50 See citation in Mandelbaum 1965: 284. In the early New York colony, Governor Edmund Andros imposed a partial form of prohibition, banning distillation except from damaged, inedible grain, because booze makers were sucking up so much of the local grain supply that people couldn't find bread (Gately 2008: 153).

51 Duke 2010.

52 Poo 1999: 127.

53 Guasch-Jané 2008.

54 Forsyth 2017: 171 (referring to the secure warehouse built to house the First Fleet's supply of rum) and 173.

55 Forsyth 2017: 37.

56 Quoted in Gately 2008: 215.

57 Gately 2008: 216.

58 Pollan 2001: 3-58.

59 Jennings and Bowser 2009.

60 In fact, Dietler 2006 notes that "most traditional forms of alcohol are made for immediate consumption: They will spoil within a few days of fermentation" (238). Most grains beers, for instance, quickly go bad without the addition of hops, an innovation that didn't occur until the ninth century in Europe.

61 Holtzman 2001.

62 Shaver and Sosis 2014. As the authors note, given the costs of kava production and consumption, it must provide significant countervailing social benefits.

63 Another historian estimated that, around the time of the French revolution, the average Parisian spent 15 percent of his or her total income on wine (quoted in Mäkelä 1983).

64 Wettlaufer et al. 2019.

65 Collaborators 2018: 12, italics added.

66 William Shakespeare, *Othello* (11, iii).

67 As Ronald Siegel notes, through the animal kingdom, individuals who get intoxicated on alcohol or plants are more prone to accidents and predation, and tend to be terrible, inattentive parents (R. Siegel 2005). A study by Sanchez et al. 2010 found that fruit that had fermented to a greater than 1 percent ethanol level ended up getting fruit-eating bats drunk, impairing their flying ability, echolocation, and communication with other bats, putting them at greater risk of injury or predation. Also see Samorini 2002: 11, 22ff.

68 As Steve Morris and colleagues have observed (Morris, Humphreys, and Reynolds 2006), despite stories of packs of elephants in the wild seeking out overripe, alcoholic fruit before embarking upon their rampages, simple physiological considerations suggest that this is impossible. Naturally alcoholic fruit is quite weak; elephants are quite large. "Extrapolating from human physiology," they observe, "a 3,000-kg elephant would require the ingestion of between 10 and 27 L of 7 percent ethanol in a short period to overly effect behavior." This is, to put it mildly, unlikely to happen in the wild. An elephant could only get drunk with access to large quantities of highly alcoholic beverages. In other words, there could be no drunken elephants without drunken people.

69 Carrigan 2020; Carrigan et al. 2014; also see Gochman, Brown, and Dominy 2016 for evidence that ADH4 in two primates, the aye-aye and the wonderfully named slow loris, allows them to make preferential use of fruit with a high alcohol content. Hockings and Dunbar speculate that it was AHD4 that allowed ape lineages that possessed it to survive a mass ape extinction that occurred during the Miocene drying period (10.4 to 5 million years ago), when monkeys' distinctive ability to digest unripe fruit gave them an otherwise decisive advantage (2020: 197).

70 See especially Hagen, Roulette, and Sullivan 2013; E. Hagen and Tushingham 2019; Sullivan, Hagen, and Hammerstein 2008.

71 It is worth noting that our closest relatives, gorillas and chimpanzees, appear to also enjoy partaking of plant-based intoxicants. Indeed, some local cultures claim to have discovered the psychoactive qualities of plants in their environment by observing non-human apes using them (Samorini 2002).

72 On lactose tolerance, see Gerbault et al. 2011; for Tibet, see Lu et al. 2016; for adaptations to diving, see Ilardo et al. 2018.

73 See Gibbons 2013 for a report on work documenting the path-dependence-bound problems with the human foot, ankle, and back.

74 The drug, disulfiram, mimics the effects of an inefficient ALDH enzyme by directly suppressing ALDH activity in the body. See Oroszi and Goldman 2004.

75 G. S. Peng et al. 2014; Y. Peng et al. 2010.

76 Goldman and Enoch 1990.

77 Park et al. 2014, Han et al. 2007. Also see Polimanti and Gelernter 2017, who argue that "the selection signatures at the ADH1B locus [the hyper-efficient variant of ADH] are primarily related to effects other than those of alcohol

metabolism," instead likely reflecting a response to the problem of infectious diseases.

78 It is worth noting that Carrigan et al. 2014 see the combination of these two enzymes as "an early stage of adaptation" to the novel problem of alcohol created by large-scale agriculture. It is an open question of how fast we should expect this particular silver bullet to spread in the absence of countervailing adaptive forces. Given the rapidity of genetic evolution and the seriousness of the alcohol problem, however, it is reasonable to assume the sluggish speed of this gene-complex spread reflects the force of countervailing pressures of the sort we will explore below: the adaptive benefits, for the individual and group, of alcoholic intoxication. In any case, the cultural evolutionary considerations explored next reinforce the claim that the Asian flushing genetic combination has been mysteriously slow to spread.

79 Frye 2005: 67.

80 Forsyth 2017: 121.

81 Gately 2008: 63.

82 Forsyth 2017: 127.

83 *Book of Odes* #220 and #255, translated by and quoted in Kwong 2013: 46. After a long description of drunken, chaotic behavior, the Ode concludes, "Drinking wine is very auspicious / Provided it is done with decency" (Waley 1996: 208, translation modified).

84 Cited in Chan 2013: 16. As Robert Eno has noted, the first mentions in our written sources of the Mandate of Heaven related to protecting the new royal line from drunkenness, which is viewed as the central vice leading to the fall of the Shang (Eno 2009: 101).

85 On early Chinese concerns about alcohol, see Poo 1999 and Sterckx 2006: 37-40.

86 Chan 2013: 16.

87 Poo 1999: ftn 23.

88 For instance, in 207 CE, the Eastern Han Grand Chancellor Cao Cao (155 to 220) announced an edict prohibiting alcohol consumption, worried that excessive drinking was leading to social chaos and endangering the state. It is worth noting that Cao Cao and his court were exempt from this ban, and in fact were famous for establishing poetic tropes centered on wine-fueled symposia (N. M. Williams 2013).

89 From James Davidson, quoted in Chrzan 2013: 20.

90 Tlusty 2001: 71.

91 From 1898, quoted in Edwards 2000: 45.

92 Quoted in Hall 2005: 79.

93 Hall 2005: 79.

94 Sherratt 2005: 21.

95 Matthee 2014: 101. Mark Forsyth also provides a helpful account of the contradictory attitudes toward alcohol in Islam (Forsyth 2017: 104-119).

96 Quoted in Matthee 2014: 100.

97 Fuller 1995.

98 Fuller 2000: 113; also see Fuller 1995: 497-498.

99 Sherratt 2005: 23.

100 Poo 1999: 135.

101 Ode #279, "Abundant Year," quoted in Kwong 2013: 46.

102 Poo 1999.

103   Chrzan 2013: 34-39.
104   Edwards 2000: 22-23. Also see T. Wilson 2005.

CHAPTER TWO: LEAVING THE DOOR OPEN FOR DIONYSUS

1   See Henrich 2015: Ch. 2 on the limits of individual human intelligence and Ch. 3 on the misadventures of stranded European explorers, stories that illustrate the helplessness of humans trying to survive without the benefit of cultural knowledge.

2   For a similarly illuminating take on historical disasters, see Christakis 2019: Ch. 2 on accounts of shipwreck survivors, where success is ultimately dependent on effective cooperation and subordinating individual needs to that of the group. It is worth noting that Christakis singles out the presence of alcohol as a causal factor in why such "unintentional communities" fail (2019: 50, 95, 99), which on the face of it would seem to contradict a major thesis of this book, which is that alcohol has helped humans ramp up the scale of cooperation. In fact, these examples actually reinforce an argument that I will be making in Chapter Five: In the absence of any cultural or ritual norms governing its use, distilled liquor (an evolutionarily novel and unusually dangerous form of alcohol, and almost exclusively what is left to shipwreck survivors) is often more harmful than beneficial for both social groups and individuals.

3   See Boyd, Richerson, and Henrich 2011; Laland 2000.

4   Again, the most readable, helpful introduction to this theme is Henrich 2015. See especially Chs. 15 ("When We Crossed the Rubicon"), 16 ("Why Us?"), and 17 ("A New Kind of Animal"). Also see Boyd, Richerson, and Henrich 2011.

5   Wrangham 2009. Wrangham believes this adaptation to fire goes as far back as *Homo erectus,* but there is some debate about this.

6   Hrdy 2009, Ch. 1 ("Apes on a Plane").

7   Haidt, Seder, and Kesebir 2008.

8   See Marino 2017 for a review.

9   See Heidt 2020 for an account of recent research on the topic.

10   Dally, Emery, and Clayton 2006; Emery and Clayton 2004.

11   B. Wilson, Mackintosh, and Boakes 1985. Although birds followed a very different evolutionary trajectory than primates, through a process of convergent evolution, corvids appear to have developed a brain region, the nidopallium caudolaterale (NCL), that is analogous in function to the prefrontal cortex (PFC) in humans, the seat of abstract reasoning and executive function (Veit and Nieder 2013). As we'll see, the PFC is a major player in any story about the adaptive function of intoxication.

12   Heinrich 1995.

13   Gopnik et al. 2017.

14   Another species that experiences menopause is orcas, probably for quite similar reasons as humans—intensive investment in the rearing of juveniles. See Fox, Muthukrishna, and Shultz 2017.

15   Sophocles 1949, lines 173-181.

16   Huizinga 1955: 108.

17   Richerson and Boyd 2005.

8   Huizinga 1955: 110.
9   *Hole.*
0   Gopnik et al. 2017, see article for supporting references.
1   Adapted from Gopnik et al. 2017, figure 2.
2   Adapted from Sowell et al. 2002, figures 3 and 4b. The gray and white matter density reflected the volume of gray and white matter as a proportion of total intercranial volume.
3   As Gopnik and colleagues observe, "Strong frontal control has costs for exploration and learning. Interference with prefrontal control areas through transcranial direct current stimulation leads to a wider range of responses on a 'divergent thinking' task, and during learning there is a characteristic release of frontal control." See citations there, especially Thompson-Schill, Ramscar, and Chrysikou 2009 and E. G. Chrysikou et al. 2013. Also see Chrysikou 2019.
4   Limb and Braun 2008.
5   Chrysikou et al. 2013. Also see a more recent study, Hertenstein et al. 2019, where transcranial deactivation of the left PFC and stimulation of the right PFC resulted in better performance on a variety of creativity and lateral thinking tasks.
6   Brown 2009: 55.
7   Brown 2009: 33
8   Brown 2009: 44.
9   Zabelina and Robinson 2010.
0   Henrich 2015.
1   Muthukrishna et al. 2018.
2   They add: "Individuals connected in collective brains, selectively transmitting and learning information, often well outside their conscious awareness, can produce complex designs without the need for a designer—just as natural selection does in genetic evolution. The processes of cumulative cultural evolution result in technologies and techniques that no single individual could recreate in their lifetime, and do not require its beneficiaries to understand how and why they work" (Muthukrishna et al. 2018).
33   Henrich 2015: 97-99.
34   Henrich 2015: 97-99, and op cit.
35   Kline and Boyd 2010, Bettencourt and West 2010.
36   Henrich 2015: Ch. 15.
37   Adapted from Muthukrishna et al. 2018, figure 9 (CC-BY).
38   Gopnik 2009: 123.
39   Gopnik 2009: 115-119.
40   Gopnik 2009: 95-95, 105.
41   Skyrms 2004: xi; cf. Yanai and Lercher 2016.
42   Dawkins 1976/2006. Indeed, sexual reproduction itself represents a kind of cooperative agreement: As long as the process by which particular genes get selected to hitch a ride on the lifeboat of the sex cell (sperm or egg) is random and thereby fair, everyone agrees to a situation where half of them will never make it. There are therefore powerful selection pressures ensuring that the selection *does* remain fair, resisting the pressure of various cheating mechanisms that strive to favor one set of genes over another.
43   For an excellent introduction to Darwinian processes working at all levels of co-

operation, from genes to cells to individuals in groups, see D. S. Wilson 2007.

44 See Hauert et al. 2002 for a review. As they note, "the diversity of names underlines the ubiquity of the issue" (1129).

45 To take a political example, the 2020 Democratic primary featured a variant of the Prisoner's Dilemma. The public good for the moderate wing of the Democratic Party clearly required uniting around a consensus candidate, but—at least until Super Tuesday—no individual moderate was willing to sacrifice their candidacy in the absence of a commitment from other rivals to coalesce around a consensus choice. Oil cartels, like OPEC and its aligned countries, are vulnerable to a rogue member breaking ranks to ramp up production at the expense of other members. At the time of writing (March 2020), in the wake of plummeting oil prices caused by the Covid-19 outbreak, Saudi Arabia appears to have decided to defect at the expense of Russia.

46 Damasio 1994; R. Frank 1988, 2001; Haidt 2001.

47 R. Frank 1988.

48 The philosopher who has done the most to redirect our attention to these sorts of relationships, and to recognize the deeper background of trust behind all human interactions, is Annette Baier. See especially Baier 1994.

49 Spinka, Newberry, and Bekoff 2001; Brown 2009: 181.

50 Brown 2009: 31-32.

51 Gopnik 2009: 11.

52 To their credit, Gopnik and her collaborators recognize this issue, noting, "Young children are rarely the source of complex technical innovations; actually designing and producing an effective tool, for example, is a challenging task that requires both innovation and executive skill." They still, however, see literal youthfulness as the key to cultural innovation: "Innovations that are effortful and rare when they first appear within a generation can become effortlessly and widely adopted by the next generation. In fact, among nonhuman animals, cultural innovations are often first produced, adopted, and spread by juveniles" (2017: 55–58).

53 Matthew 18:3; *Daodejing* Chs. 10, 20, 28, 55.

54 Braun 1996: 40.

55 Braun 1996: 14.

56 An excellent general introduction to the physiological effects of alcohol is provided by Sher and Wood 2005; Sher et al. 2005.

57 Olive et al. 2001, Gianoulakis 2004.

58 It is worth noting that, while alcohol is sometimes also recommended as a sleep aid, this is not the best advice. Its sedative effect, through inhibition of brain activity, does make it easier to initially fall asleep. However, the brain is always trying to adapt and regain equilibrium, and there is evidence that it responds to the inhibitory effects of alcohol by boosting the excitatory system, leading to a rebound effect. This is why alcohol-induced sleep is often deep and quick in onset, but causes one to wake up in the middle of the night, and then have trouble getting back into deep sleep.

59 Olsen et al. 2007.

60 Miller and Cohen 2001.

61 Mountain and Snow 1993.

62 Heaton et al. 1993; Lyvers and Tobias-Webb 2010; Nelson et al. 2011; Easdon et al. 2005. Lyvers, Mathieson, and Edwards 2015 have similarly shown that alcohol nega-

tively effects performance in another experimental task, the Iowa Gambling Task (IGT), which more specifically depends on the function of the ventromedial prefrontal cortex (VMPFC).

53 Nie et al. 2004.

54 Steele and Josephs 1990 were the early champions of the "myopia" theory of alcohol intoxication; also see Sayette 2009; Sher et al. 2005: 92ff; and Bègue et al. 2013 for reviews of the literature.

55 Dry et al. 2012.

56 On disinhibition, see Hirsch, Galinsky, and Zhong 2011; on impairment of PFC and ACC and general cognitive control, see Lyvers 2000, Curtin et al. 2001, Hull and Slone 2004. Malcolm Gladwell (Gladwell 2010) presents the disinhibition and myopia accounts as competing, whereas it seems more likely that they are complementary, simply two aspects of alcohol's downregulation of the PFC and associated systems.

57 Easdon et al. 2005.

58 Carhart-Harris et al. 2012, 2014; Kometer et al. 2015; Pollan 2018: 303–5. A study by Dominguez-Clave et al. 2016 suggests that ayahuasca similarly relaxes top-down constraints and "the cognitive grip exerted by the frontal cortex."

59 A. Dietrich 2003.

70 Kuhn and Swartzwelder 1998: 181.

71 This tension was most prominently and explicitly identified by the German philosopher Friedrich Nietzsche. See, especially, *The Birth of Tragedy* (1872).

72 *Laws* II, citation from Szaif 2019: 107.

73 Huxley 1954/2009: 77.

74 Fertel 2015.

CHAPTER THREE: INTOXICATION, ECSTASY, AND THE ORIGINS OF CIVILIZATION

1 Braidwood et al. 1953; Dietler 2006; Hayden, Canuel, and Shanse 2013; Katz and Voight 1986. But also see Dominy 2015 for an opposing view.

2 Arranz-Otaegui et al. 2018.

3 Hayden, Canuel, and Shanse 2013; Arrenze-Otaegui et al. 2018.

4 For instance, a clay seal from a site in northern Iraq, dating to perhaps 4000 BCE, shows two people drinking with straws from a large jar, which certainly contains more than water. Sumerian beer was a serious form of what we now call "unfiltered" beer: The yeast was left in the jar during fermentation and formed a solid crust on the surface of the beer, which was then accessed by poking straws through the yeast cake (Katz and Voight 1986).

5 Fay and Benavides 2005. Ninety-five percent of wine yeasts are closely related, suggesting a single origin of grape wine production, probably in Mesopotamia, and then subsequent spread through the Middle East and Europe (Sicard and Legras 2011).

6 There is strong evidence for widespread production of *chicha* throughout South America from the early centuries of common era (Jennings and Bowser 2009).

7 P. L. Watson, Luanratana, and Griffin 1983.

8 See discussion in Carmody et al. 2018.

9   Friedrich Hölderlin, "Dichterberuf," *Sämtliche Werke.* 6 Bände, Band 2, Stuttgart 1953, S. 46-49, translation by author. Accessed through an open-source website: http://www.zeno.org/Literatur/M/H%C3%B6lderlin,+Friedrich/Gedichte/Gedichte+1800-1804/%5BOden%5D/Dichterberuf.

10 Quoted in Mattice 2011: 247.

11 Cited in Kwong 2013: 56, translation modified. Perhaps my personal favorite of such wine-inspired poetry is that of Tao Qian or Tao Yuanming (365-427). See Mattice 2011, Kwong 2013, and Ing, In preparation: Ch. 3 for excellent surveys of the role of wine in ancient Chinese and Greek culture.

12 Quoted in Gately 2008: 16.

13 Gately 2008: 56.

14 Roth 2005: 122.

15 Quoted in Roth 2005: 108. See Roth for an excellent review of the role of intoxicants in ancient and contemporary creative culture, as well as Djos 2010.

16 Lebot, Lindstrom, and Merlin 1992: 155.

17 See the comments of jazz clarinetist Mezz Mezzrow (1899-1972), a cannabis guru in the American counterculture of the mid-twentieth century, quoted in Roth 2005: 130.

18 Eliade 1964 is the classic work on shamanism in a cross-cultural context. For a more recent survey, framed in a modern evolutionary perspective, see Winkelman 2002.

19 Lietava 1992; although see Sommer 1999 for an argument that the plant traces were introduced after the burial by local wildlife.

20 Harkins 2006.

21 Amabile 1979; also see Harkins 2006 on the "evaluation-performance relationship."

22 See Aiello et al. 2012 and the literature reviewed there, as well as Beilock 2010; DeCaro et al. 2011.

23 Gable, Hopper, and Schooler 2019; Mooneyham and Schooler 2013.

24 Haarmann et al. 2012.

25 Quoted in Katz and Voight 1986.

26 Proverbs 31:6.

27 Quoted in Sayette 1999.

28 Cited in Kwong 2013: 52 (translation modified).

29 See Patrick 1952: 45-47 for a survey of early functional theories, almost all of which center on escape from reality or anxiety relief.

30 Horton 1943: 223.

31 See, e.g., Sher et al. 2005: 88, who also emphasize the importance of "expectancy effects," or the power that alcohol can have purely as a placebo in the presence of certain cultural expectations.

32 See especially O. Dietrich et al. 2012; Dunbar 2017; Dunbar et al. 2016; Wadley 2016; Wadley and Hayden 2015; all of these views will be discussed more below.

33 Nagaraja and Jeganathan 2003.

34 Adapted from Nagaraja and Jeganathan 2003, figure 2.

35 Levenson et al. 1980.

36 Coined by Levenson et al. 1980; cf. Baum-Baicker 1985, Peele and Brodsky 2000, Müller and Schumann 2011.

37 As Sayette 1999 notes, the myopia-inducing effect of alcohol causes the drinker to focus on immediate surroundings, which means that stress reduction works best in

combination with pleasant distractions, especially social interactions. This is why solitary drinking may lead to mostly negative outcomes, a topic to which we'll return in Chapter Five.

8 Bahi 2013; also see literature reviewed there.

9 Relevant to this point is MacAndrew and Edgerton's (1969) argument that alcohol has historically functioned to facilitate a "Time Out" for individuals that allows them to assert their desire for individual freedom against social constraints and demands.

0 See Tooby and Cosmides 2008 for the classic argument concerning the need for an evolved capacity to evaluate trustworthiness in others.

1 Willis and Todorov 2006; van 't Wout and Sanfey 2008; Todorov, Pakrashi, and Oosterhof 2009.

2 Cogsdill et al. 2014, which also contains a review of the relevant literature.

3 R. Frank, Gilovich, and Regan 1993.

4 See Sparks, Burleigh, and Barclay 2016 for a review and citations of relevant literature. In one nice study, David DeSteno and colleagues zeroed in on a set of specific and predictive nonverbal cues to which people attend when judging someone as an untrustworthy partner in an economic game: hand touching, face touching, crossing the arms, and leaning away (DeSteno et al. 2012). We at least implicitly recognize that people who fidget too much are thinking too much, and indeed it was the hand-touching arm-crossers who tended to defect in economic trust games. In a wonderful twist that allowed them to filter out other potential confounds, they found that people also distrusted a robot named Nexi when they caused it to perform the same behavioral cues. They didn't dislike the arm-crossing Nexi more; they simply wouldn't trust it with their money.

5 Darwin 1872/1998; Ekman 2006; also cf. Bradbury and Vehrencamp 2000. In China, early (roughly 300 BCE) Confucian thinkers saw emotional displays, read through the "countenance" (si 色), tone of voice or pupils of the eyes, as the most reliable way to judge the genuine moral state of others (Slingerland 2008a).

6 Tracy and Robbins 2008.

7 Ekman and O'Sullivan 1991; M. G. Frank and Ekman 1997; Porter et al. 2011; ten Brinke, Porter, and Baker 2012; Hurley and Frank 2011.

8 On smiles, see Ekman and Friesen 1982, Schmidt et al. 2006; on laughter, see Bryant and Aktipis 2014.

9 Centorrino et al. 2015; also see Krumhuber et al. 2007, where experimenters were able to create dynamic "fake" versus "authentic" smiles in supposed trust game partners, and found that 60 percent of subjects chose to play with an authentically smiling counterpart, 33.3 percent with a fake smiler, and 6.25 percent with a partner displaying a neutral expression. Also see Tognetti et al. 2013 for a review of evaluation of cooperation based on facial signals, especially authentic smiles; and Levine et al. 2018 on people's tendency to put greater trust in those who have displayed signals of genuine emotion.

0 Dijk et al. 2011 and op cit. Also see Feinberg, Willer, and Kaltner 2011 on embarrassment as a positive social signal of trustworthiness.

1 ten Brinke, Porter, and Baker 2012.

2 Boone and Buck 2003, and op cit.

3 Rand, Greene, and Nowak 2012; Capraro, Schulz, and Rand 2019; Rand 2019.

54   That is, with the exception of rationalist, "cold cognition" models of ethics that have been historically quite rare, but more or less took over Western philosophy a few hundred years ago. For more on spontaneity and trust, see Slingerland 2014: Ch. 7.

55   Dawkins et al. 1979.

56   For instance, work by Silk 2002 suggests that chimps use nonverbal cues, such as coughs or low grunts, to communicate trustworthiness and lack of aggression to others.

57   Byers 1997.

58   Ekman 2003: Ch. 5; also see discussion in R. Frank 1988.

59   Porter et al. 2011.

60   Michael Sayette (personal communication) observes that there is also a potential dark side to this function, in that tyrannical rulers could exploit the truth-telling function of alcohol to keep their underlings in line. In this regard, it is worth noting that Stalin apparently kept the officials under him in a constant state of terror and submission, unable to effectively scheme among themselves, by randomly summoning them, sometimes in the middle of the night, to heavy drinking sessions where Stalin himself remained sober.

61   This phrase appears in "King Cheng's Trip to Chengpu" (Ma 2012: 148), a fragmentary and challenging to decipher bamboo text that is part of the collection of Warring States (roughly 3rd c BCE) texts purchased and published by the Shanghai Museum.

62   Quoted in Forsyth 2017: 95.

63   Gately 2008: 15, also see pp. 15-16 as well as Szaif 2019 on ancient Greek views on wine and truthfulness.

64   Gately 2008: 12 on Greek oaths, Forsyth 2017: 126-127 on Viking oaths, McShane 2014 on 17th century England.

65   Lebot et al. 1992: 119.

66   Fuller 2000: 37.

67   Ode #174, translated by Waley 1996: 147.

68   Quoted in Gately 2008: 452.

69   "The beer vat, with its characteristic shape, serves as symbolic indication of social interaction on early seal impressions and other banquet scenes" (Michalowski 1994: 25).

70   Chrzan 2013: 36.

71   Heath 1990: 268.

72   Powers 2006: 148.

73   Chrzan 2013: 30-31.

74   Austin 1979: 64.

75   Fray Bernardino de Sagagún, quoted in Carod-Artal 2015.

76   Price 2002, Fatur 2019.

77   Mandelbaum 1965, Gefou-Madianou 1992.

78   As Mandelbaum notes, "Drinking is more often considered for those who grapple with the external environment than for those whose task it is to carry on and maintain a society's internal activities. This distinction was anciently symbolized in India by the difference between the god Indra, the scourge of enemies, the thunderer, the roisterer, and heavy drinker, and Varuna, the sober guardian of order and morality" (1979: 17-18).

79 Jennings and Bowser 2009, citing Weismantel 1988: 188.

80 D. Heath 1958; D. B. Heath 1994.

81 Madsen and Madsen 1979: 44.

82 Forsyth 2017: 28.

83 Jennings and Bowser 2009: 9.

84 Tlusty 2001: 1. Also see her description of the "contract drink" in early modern Germany (92-93), which "created a bond between drinkers that was weightier than a verbal promise, or even a written agreement."

85 Michalowski 1994: 35-36.

86 *Analects* 10.8. This phrase is more elegant in classical Chinese, only four characters (*wei jiu bu liang* 唯酒不量), and I've always rather fancied having it inscribed on my tombstone.

87 Quoted in Roth 2005: 55.

88 Szaif 2019.

89 Mattice 2011: 246.

90 Mars 1987.

91 Shaver and Sosis 2014.

92 Osborne 2014: 60.

93 Cited in Pollan 2001: 23; as Pollan observes, Emerson's readers would have known he was referring to the alcoholic properties of apples in calling them a "social fruit."

94 James 1902/1961.

95 Duke 2010: 266.

96 Bourguignon 1973.

97 Ehrenreich 2007: 5.

98 Radcliffe-Brown 1922/1964: 252, cited in Rappaport 1999: 226.

99 Rappaport 1999: 227.

00 Tarr, Launay, and Dunbar 2016, and see the literature review there.

01 Reddish, Bulbulia, and Fischer 2013.

02 See Earle 2014: 83 for references.

03 Forsyth 2017: 44; see his account of this festival pp. 42-49.

04 McGovern 2020: 89.

05 Reinhart 2015; cf. Allan 2007.

06 Guerra-Doce 2014: 760.

07 Fuller 2000: 28. Also see the account of *yají* use among the Barasana discussed in Tramacchi 2004.

08 See Mehr et al. 2019; the alcohol results, which did not make the final published version of the paper, are from personal communication with the lead author.

09 Pitek found that, of the 160 cultures in the "eHRAF World Cultures" database tagged with "ecstatic religious practices," only 140 of these met our criteria. This is because a previous category of "orgies" was converted to "ecstatic religious practices" in 2000, resulting in the latter category including sexual practices that do not necessarily involve ecstatic states. An additional 154 cultures not found in the initial search came up when Pitek searched for "trance," but time considerations prevented us from exploring these cultures, so the 71 percent estimate is almost certainly on the low side. Many thanks to Emily for her hard and thoughtful work on this project.

10 Fernandez 1972: 244.

111 Machin and Dunbar 2011; also see Dunbar 2017 on the role of alcohol in particular.

112 Crockett et al. 2010, Wood et al. 2006; see review in J. Siegel and Crockett 2013.

113 For modern MDMA-based religions, see St. John 2004 and Joe-Laidler, Hunt, and Moloney 2014.

114 Kometer et al. 2015.

115 Fr. Bernandino de Sahagún, *History of the Things of New Spain,* Book 10, quoted in Furst 1972: 136.

116 Furst 1972: 154-156.

117 "It was in this context of cultural disarray that the Ghost Dance religion emerged and incited widespread enthusiasm. The Ghost Dance religion prophesized the imminent dawn of a golden age of pan-Indian harmony. It emphasized the need for peace between the tribes and fashioned a sense of intertribal unity based upon their shared contempt for white man's civilization." (Fuller 200: 38ff; also see Shonle 1925.)

118 Quoted in Smith 1964.

119 L. Dietrich et al. 2019; cf. O. Dietrich et al. 2012. For a popular account of Göbekli Tepe and the connection to alcohol and the beginnings of civilization, see Curry 2017.

120 On the category and function of "high arousal" or "imagistic" rituals, see Whitehouse 2004; cf. McCauley and Lawson 2002.

121 To be fair, there is, as yet, no direct evidence, in the form of chemical residues, etc., that the vats and vessels at the site were used for alcoholic beverages. In their latest statement on the topic, Dietrich and Dietrich characterize the evidence for the use of "mind-altering beverages" at the site as "tentative" (Dietrich and Dietrich 2020: 105).

122 Duke 2010: 265, and op cit., and changing "Inka" to the more common rendering "Inca."

123 Ode #171, translation by James Legge; see discussion in Poo 1999.

124 Doughty 1979: 78-9. Also see descriptions of the public function of the *minka,* contemporary formal ritual feasting exchanges in the Andes (Bray 2009; Jennings and Bowser 2009), accounts of which extend back to the sixteenth-century Incan Empire.

125 As Dwight Heath notes of alcoholic beverages, "As economic commodities, they tend to represent relatively high value in small volume and play a variety of roles in the economics and prestige systems of societies that use them" (D. Heath 1990: 272).

126 Enright 1996.

127 Gately 2008: 4-5.

128 Nugent 2014: 128.

129 See Hockings, Ito, and Yamakoshi 2020 for an account of the use of palm wine in Guinea, West Africa, esp. p. 51.

130 See Shaver and Sosis 2014 and op cit, as well as Bott 1987: 191.

131 Greg Wadley and Brian Hayden, for instance, touch on many of the functions of intoxication surveyed in this chapter. They argue that the cultivation and production of psychoactive drugs supported the formation of large-scale societies by "1) providing a motive to adopt and maintain cultivation; 2) enhancing pro-sociality, allowing the maintenance and governance of larger, more coordinated groups; 3)

instilling acquiescence in and providing solace to subordinates; 4) enticing people into labour arrangements, enhancing their efficiency, and compelling them to keep working" (Wadley and Hayden 2015, as summarized in Wadley 2016). Also see Courtwright 2019 and Smail 2007 on the early modern economics of psychoactives and their role in creating and supporting modern societies.

32 Katz and Voight 1986; Hayden 1987; Joffe 1998; Hayden, Canuel, and Shanse 2013.

CHAPTER FOUR: INTOXICATION IN THE MODERN WORLD

1 Klatsky 2004; Braun 1996: 62-68.
2 Lang et al. 2007; Britton, Singh-Manoux, and Marmot 2004.
3 O'Connor 2020.
4 Khazan 2020.
5 Jarosz, Colflesh, and Wiley 2012.
6 Sayette, Reichle, and Schooler 2009.
7 Gately 2008: 25.
8 Gately 2008: 445; also see Moeran 2005: 38.
9 Lalander 1997, reported in Heath 2000: 185. Also see Heath (Heath 2000: 186) on beer dances among the Azande people of North Central Africa, which among other functions allow for the drunken airing of grievances without creating offense, and Dennis 1979 on the role of the drunk "truth-teller" at a Oaxacan village banquet, who is given license, because of his impaired state, to voice local complaints to a visiting dignitary.
10 Hunt 2009: 115.
11 Quoted in Andrews 2017.
12 Allen 1983.
13 Bettencourt and West 2010.
14 Marshall 1890.
15 Andrews 2017; the quotation is drawn from Dutton 1984: 11.
16 As Andrews notes, "Examples of inventions first articulated in bars are plentiful, from the first electronic digital computer and MRI machines to Discovery Channel's Shark Week. A large part of the modern computer industry emerged out of an informal group that met at The Oasis bar and grill...and several other Silicon Valley watering holes have become legendary as common meeting places for engineers during the early decades of the high tech industry."
17 Walton 2001: xiv; see his excellent account of the role of intoxication in events such as these.
18 Quoted in Roth 2005: 58.
19 James 1902/1961: 388.
20 Huxley 1954/2009: 17, 73.
21 Carhart-Harris and Friston 2019.
22 Markoff 2005; Pollan 2018.
23 Pollan 2018: 175-185; Markoff 2005: xix.
24 Florida 2002.
25 For a wealth of references see James Fadiman's website, www.jamesfadiman.com.
26 Hogan 2017.

27 2015 interview with CNN, quoted in Hogan 2017.

28 Anderson et al. 2019. Another recent study that tracked the experience of micro-dosers over six weeks (Polito and Stevenson 2019) reported an increase in creativity among active microdosers, but this was based on a self-reported sense of creativity rather than an experimental measure.

29 Prochazkova et al. 2018.

30 "Estimating drug harms: A risky business," Centre for Crime and Justice Studies, Briefing 10, October 2009.

31 Pollan 2018: 318-9. Also see his comments on the importance of "neural diversity" (17) and the idea that the plant toxins that drive psychedelic effects might function like a "kind of cultural mutagen, not unlike the effect of radiation on the genome" (149).

32 Samorini 2002. Samorini, in turn, was inspired by earlier work by Edward de Bono in the 1960s on intoxicants as "depatterning tools" used in order to shake up ordi-nary thought. Reflecting the tension we have characterized as that of Apollo vs. Dionysus, de Bono writes, "The function of language is to reinforce existing mod-els; the function of [intoxication] is to facilitate escape from these models" (de Bono 1965: 208, quoted in Samorini 2002: 85).

33 T. Leary 2008, quoted in Joe-Laidler, Hunt, and Moloney 2014: 63.

34 Noted in "On the road again: Companies are spending more on sending their staff out to win deals" 2015: 62.

35 L. ten Brinke, Vohs, and Carney 2016.

36 Giancola 2002; see Hirsch, Galinsky, and Zhong 2011 on general topic of intoxica-tion and disinhibition.

37 "To your good stealth: A beery club of Euro-spies that never spilt secrets" 2020.

38 It is worth mentioning one other multinational research collaboration hatched in Vancouver that was apparently catalyzed by alcohol. Richard Beamish, now an emeritus fisheries researcher in British Columbia, is the organizer of a massive col-laboration between Canadian, American, and Russian researchers to study the migratory patterns of Pacific salmon. A newspaper account of the project (C. Wil-son 2019) traces its genesis to a vodka-drinking session after a Vancouver workshop, where Beamish and a Russian scientist, PFCs sufficiently downregulated, started speculating on the possibilities of such a multinational effort. "Just for fun," Beam-ish reportedly declared, presumably hovering around at least 0.08 BAC, "How about if I just arrange it?" The result was an award-winning project that has transformed our understanding of an important fishery.

39 "The 90 percent economy that lockdowns will leave behind," *The Economist*, April 30, 2020.

40 While finalizing this chapter draft I came across a short piece by Carl Benedikt Frey (Frey 2020) that makes similar predictions about the chilling effect of Covid-19 on innovation, primarily through its disruption of in-person socializing. He cites a re-cent study that suggest that physical interaction is crucial to the establishing of research collaborations (Boudreau et al. 2017), as well as a fascinating one that took advantage of another natural experiment—the last-minute cancellation of the 2012 American Political Science Association Annual Meeting in New Orleans caused by Hurricane Isaac—to show that the elimination of such in-person contact leads to a significant decrease in the subsequent co-authorship of articles (Campos, Leon,

and McQuillin 2018). Although neither of these studies explicitly mentions the role of alcohol, we can venture with some confidence that the missed socializing activities at the conference in New Orleans would have been well lubricated.

41 Her clinical summary of experimental work on alcohol and social interactions should sound familiar to anyone who has ever been to a cocktail reception, office party, or post-work pub session: "Reports in the literature have explained [enhanced] conviviality by noting that conversations appear to flow with greater facility; that persons exhibit raised spirits; and that at low doses of ethanol there is a greater degree of social interaction. People under low and/or moderate doses of alcohol have been described as more talkative. Voice tones have been described as louder and more boisterous and of a higher pitch" (Baum-Baicker 1985: 311).

42 See especially Hull and Slone 2004 and the literature reviewed in Peele and Brodsky 2000 and Müller and Schumann 2011.

43 Chrzan 2013: 137.

44 E.g., Horton 1943.

45 As Sayette et al. 2012 observe in the introduction to their study, "Given the widespread use of alcohol in social situations, it is notable that both alcohol researchers and social psychologists have generally neglected the effects of alcohol on social bonding."

46 Sayette et al. 2012.

47 Sayette et al. 2012, figure 1; permission to reproduce secured from Sage Publication via the Copyright Clearance Center (license #4947120323569).

48 Kirchner et al. 2006.

49 Fairbairn et al. 2015.

50 Orehek et al. 2020: 110–111, with citations removed for readability; please refer to the study for an excellent literature review on this topic. It is worth noting that the study performed by Orehek et al. looked at observers' ratings of the personalities of subjects shown in video clips, finding evidence that alcohol increased the positivity of the observers' ratings but not the accuracy.

51 Ban Gu, *History of the Han*. Also see Mattice 2011: 246–47 on the importance of literati drinking parties in traditional Chinese culture.

52 On mead halls, see Gately 2008: 55; on English alehouses, see Martin 2006: 98.

53 Chrzan 2013: 65, op cit.

54 See Martin 2006: 195 on the *kabak* in late-nineteenth and early-twentieth centuries Russia, which functioned much like the English alehouse or pub. "More than a club and a library, more than a drinking hole, the *kabak* was 'the centre of village public life'" (2006: 195, op cit.).

55 As Stuart Walton puts it, "For most it is the evening's first glass of wine, bottle of beer or G&T that most cheeringly announces the spiritual function of intoxication, its role, coeval with humanity itself, in displacing the given world by transformation of the way the brain works" (Walton 2001: 129).

56 Mass Observation 1943, quoted in Edwards 2000: 28–29.

57 A popular piece by Dunbar (Dunbar 2018) provides a readable account of the results from the study reported in Dunbar et al. 2016.

58 Dunbar 2017.

59 Dunbar and Hockings 2020: 1.

60　"Last Orders for Political Drinking, Waning Interest in Booze Is Transforming British Politics," Bagehot column, *The Economist,* June 2, 2018.

61　Peele and Brodsky 2000.

62　Rogers 2014: 163.

63　Hart 1930: 126, from Patrick 1952: 46.

64　Quoted in Walton 2001: 22.

65　He then goes on to compare the enhanced sociality of intoxicated adults at a social gathering to the sort of deep spiritual connection that accompanies truly human lovemaking, rather than merely animalistic coupling. As Robin Osborne explains, "To be at a party and not intoxicated...is like having sex without feeling some spiritual attraction to the sexual partner. Not to be intoxicated, on this analogy, is the equivalent of being immature: intoxication and proper adult passion go together" (Osborne 2014: 41).

66　See W. H. George and Stoner 2000 for a review.

67　See Sher et al. 2005 for a review.

68　Lee et al. 2008, Sher et al. 2005.

69　*Macbeth,* Act 2, Scene 3.

70　Quoted in Roth 2005: 52.

71　Lyvers et al. 2011, and Chen et al. 2014; although cf. null results in Maynard et al. 2015.

72　Dolder et al. 2016, and op cit.

73　J. Taylor, Fulop, and Green 1999.

74　Quoted in Bègue et al. 2013.

75　Van den Abbeele et al. 2015.

76　Reproduced with the permission and assistance of the photographer; many thanks for his generosity.

77　Bègue et al. 2013. The follow-up study in this report, using a balanced placebo design, found that this self-enhancement effect was also produced by alcohol expectancy, not merely actual alcohol consumption.

78　Hull et al. 1983.

79　Banaji and Steele 1988, reported in Banaji and Steele 1989.

80　Quoted in Roth 2015: 8.

81　Müller and Schumann 2011.

82　From Ing, in preparation, translation by Ing, both translation and quotation from Ing slightly modified to render *jiu* 酒 as "wine" rather than "ale." Neither is ideal, but "wine" is more neutral and, in my view, the best compromise between readability and accuracy.

83　As Peele and Brodsky summarize: "Sociability is often mentioned in surveys as a primary motive for and consequence of drinking. In a diary study among young adults in Australia, the top two reasons listed by both men and women for drinking were to be sociable (30-49 percent) and to celebrate (19-15 percent). In a questionnaire survey in four Scandinavian countries, the positive consequences of drinking were 'manifested first and foremost by a loss of inhibitions in company with other people and being better able to establish contact with other people.' A survey of French-Canadians found conviviality the most prevalent (64 percent) perceived benefit of alcohol" (Peele and Brodsky 2000 and op cit.).

84　As Müller and Schumann (2011) conclude in their review of the extensive empirical literature on the topic, "Alcohol reduces social inhibition, discomfort in social situations, and social anxiety; increases talkativeness; and increases the tendency to talk about private affairs."

85 Müller and Schumann 2011; cf. Booth and Hasking 2009.

86 Meade Eggleston, Woolaway-Bickel, and Schmidt 2004; Young et al. 2015.

87 Bershad et al. 2015; Dolder et al. 2016.

88 See the literature reviewed in Baum-Baicker 1985 and Müller and Schumann 2011.

89 Wheal and Kotler 2017: 14-15.

90 Nezlek, Pilkington, and Bilbro 1994: 350.

91 Turner 2009.

92 Hogan 2017; cf. Wheal and Kotler's discussion of "ecstatic technologies" for achieving "group flow" (Wheal and Kotler 2017: 2ff).

93 Ehrenreich 2007: 163.

94 Ehrenreich 2007: 21-22.

95 Haidt, Seder, and Kesebir 2008.

96 Durkheim 1915/1965: 428. To be fair, Durkheim elsewhere does acknowledge the "ritual use of intoxicating liquors" (248), but chemical intoxicants play a very small role in his account of human ritual and bonding.

97 Rappaport 1999: 202.

98 Forstmann et al. 2020. It should be noted that alcohol, although far and away the most popular psychoactive consumed (80 percent), produced the lowest reports of transformative experiences, below even no substances at all. As the authors note, this may be attributable to the curve of alcohol's effects, which is positive in the moment but leads to unpleasant physiological consequences the day after. In other words, unlike other drugs, alcohol typically leaves you with a crushing hangover (Supplementary materials, pp. 8-9). The issue merits further study, though, as does the possibility that the same sorts of people who refrain from drug use at these events also tend to sit on the sidelines when the singing and synchronous dancing begins.

99 Nemeth et al. 2011.

100 "Man, being reasonable, must get drunk; / The best of life is but intoxication: / Glory, the grape, love, gold, in these are sunk / The hopes of all men, and of every nation; / Without their sap, how branchless were the trunk / Of life's strange tree, so fruitful on occasion! / But to return—Get very drunk; and when / You wake with head-ache, you shall see what then." Stanza 179, Canto II, *Don Juan* (1819–1824).

101 For an excellent piece on Nietzsche and Dionysian ecstasy, which directed me to many of the passages cited below, see Luyster 2001.

102 Nietzsche 1872/1967: 37.

103 Nietzsche 1891/1961: 207.

104 Cited in N. M. Williams 2013.

105 Cited in Williams 2013; also see Kwong 2013 and Ing, in preparation, on the trope of wine-fueled ecstasy in traditional Chinese poetry.

106 Ing, in preparation.

107 See youtu.be/yYXoCHLqr4o; for more on animals and intoxicants, see Samorini 2002.

108 R. Siegel 2005: 10.

109 Camus 1955: 38.

110 M. Leary 2004: 46. On this topic, also see Baumeister 1991.

111 Huxley 1954/2009: 63.

112 Quoted in Walton 2001: 119.

113 Good Friday Experiment (Pahnke 1963), with a six-month follow-up, and most of the subjects later tracked down for a twenty-plus year follow-up (Doblin 1991). See Joe-Laidler, Hunt, and Moloney 2014 for survey data on MDMA users and the reported positive effect of the "drug self" on the "everyday self," and MacLean, Johnson, and Griffiths 2011; Rucker, Iliff, and Nutt 2018; Studerus, Gamma, and Vollenweider 2010 for more recent experimental work involving psychedelics and spiritual and mental well-being. Pollan 2018 also provides an excellent, readable overview of this body of research.

114 Griffiths et al. 2011.

115 Anderson et al. 2019; Dominguez-Clave et al. 2016.

116 "Psychedelic Tourism Is a Niche But Growing Market," *The Economist*, International, June 8, 2019.

117 Talin and Sanabria 2017.

118 Sharon 1972: 131.

119 Walton 2001: 133.

120 Quoted in Walton 2001: 256.

121 See quotation from William Booth, founder of the Salvation Army, in Chapter One.

122 Walton 2001, recently re-issued as *Intoxicology: A Cultural History of Drink and Drugs* (2016).

123 Walton 2001: xvii. In a similar vein, Walton observes, "Suicide may be tragic or enraging but scarcely in itself an evil act, while auto-eroticism is positively urged upon us a beneficial to health. And yet intoxication can't be persuaded into the light. It remains, even in the face of its virtual universality, something that we have to pretend we don't do, or at least not deliberately, or at least not often, or at least only after we have done a decent day's work" (2001: 46).

124 Heath 2000: 67.

125 Quoted in Roth 2005: xiii.

126 Walton 2001: 204. Cf. pp. 234-5: "We make a fundamental mistake in seeing intoxication as a sad substitute for real fulfillment, instead of what it simply and irreducibly is—an integral component of a life fully lived. There may be higher things to dwell on, in the way of fine art or true love or transports of the soul, but they are not defeated by intoxication, and anyway they don't show their faces half often enough."

127 Specifically, "Doxology C," usually attributed to Arius Didymus (dates unclear, 1st—3rd c. CE).

128 Szaif 2019: 98.

129 Quoted in Boseley 2018.

130 Müller and Schumann 2011.

131 "Epidemiological data show...that the majority of people who consume psychoactive drugs with an addiction potential are not addicts and will never become addicted...Of those people who are classified as current alcohol drinkers in the United States, 14.9 percent are diagnosed as addicts based on the SAMHSA (2005) report...In the European Union, about 7.1 percent of the daily drinkers of alcohol are alcohol dependent...From surveys of this kind, it is clear that the majority of

psychoactive drug users are not and will never be drug addicts" (Müller and Schumann 2011, and op cit.). Also see, however, Grant et al. 2015, discussed below in Chapter Five, whose work suggests the more widespread presence of "mild" alcohol use disorder.

32 Eliade 1964: 223, 401. He claims that the use of chemical intoxicants in shamanistic practices "is a recent innovation and points to a decadence in shamanistic technique. Narcotic intoxication is called on to provide an *imitation* of a state that the shaman is no longer capable of attaining otherwise" (401). This so flies in the face of extensive archaeological evidence, reviewed above, that psychedelics have played a role in shamanistic practices from the very beginning that one has to credit these statements to a very powerful prejudice. As one commentator rather sharply observes, Eliade's dismissal of chemical intoxicants is not based on any specific scholarship, but rather his "bourgeois aversion to intoxication in relation to the religious life" (Rudgley 1993: 38).

33 Roth 2005: xix.

34 Huxley, "Drugs That Shape Men's Minds," included in 1954/2009: 14.

35 Huxley, "Heaven and Hell," included in 1954/2009: 155.

36 Cited in Kwong 2013 (translation modified).

37 Kwong 2013.

38 Baudelaire 1869, translation by author.

CHAPTER FIVE: THE DARK SIDE OF DIONYSUS

1 One of the earliest Chinese dictionaries, dating to the first century CE, glosses the word we've been rendering as "wine" (*jiu*), but which broadly refers to all alcoholic beverages, as follows: "Wine/alcohol (*jiu* 酒) means 'to achieve' (*jiu* 就). It is what is used to achieve the good and the bad in human nature." Chinese lexicographers love to define words in terms of homophones. In this case, as Nicholas Williams notes, this rhyming gloss "introduces the duality of alcohol. It is something like a catalyst that can help to realize both the positive and negative potential of human beings" (N. M. Williams 2013).

2 *Genesis* 5:20.

3 Quoted in Forsyth 2017: 144-145.

4 Heath 1976: 43.

5 Heath 2000.

6 Organization 2018.

7 www.niaaa.nih.gov/publications/brochures-and-fact-sheets/alcohol-facts-and -statistics#:~:text=Alcohol%2DRelated%20Deaths%3A,poor%20diet%20and%20 physical%20inactivity.

8 Lutz 1922:105, quoted in Mandelbaum 1965.

9 Grant et al. 2015. "Mild" alcohol use disorder was defined as the presence of two to three symptoms of AUD on the 2103 revised *Diagnostic and Statistical Manual of Mental Disorders (DSM-5)*, which include answering in the affirmative to questions such as, in the last year, have you "Had times when you ended up drinking more, or longer, than you intended?" or "More than once wanted to cut down or stop drinking, or tried to, but couldn't?"

10   See George Koob's work on "allostasis" (Koob 2003; Koob and Le Moal 2008).

11   Sher and Wood 2005; Schuckit 2014.

12   Sher and Wood 2005.

13   An excellent popular account of this research can be found in Yong 2018.

14   For a good introduction to Southern vs. Northern drinking cultures, see the description of Ruth Engs's work in Chrzan 2013: 39-41.

15   Lemmert 1991. It is also worth noting that, in the U.S., Jewish alcoholism rates are also much lower than the national average, which likely reflect the integration of wine into both meals and regular religious rituals in the home (Glassner 1991).

16   "On average, the ABV for beer is 4.5 percent; for wine, 11.6 percent; and for liquor, 37 percent, according to William Kerr, senior scientist at the Alcohol Research Group of the Public Health Institute" (Bryner 2010).

17   Rogers 2014: 84.

18   See Rogers 2014: 84-93 for a readable introduction to the history of distillation.

19   Gately 2008: 71–72.

20   Kwong 2013, ftn. 32.

21   Smail 2007: 186.

22   Edwards 2000: 38-39.

23   Edwards 2000: 197.

24   *I Li,* "Drinking Ceremony of the Countryside," cited in Poo 1999.

25   Schaberg 2001: 230; also see 228–29.

26   From the *Shi Ji,* quoted in Poo 1999: 138.

27   Quoted in Fuller 2000: 30.

28   Mars and Altman 1987: 272.

29   Chris Kavanaugh, personal communication.

30   Heath 1987: 49.

31   Doughty 1979: 67; cf. Mars and Altman on the Georgian context: "It is unheard of for men to drink alone—the role of wine is essentially social, formalized and specific to the feast" (1987: 275).

32   Toren 1988: 704.

33   See research reported in Lebot, Lindstrom, and Merlin 1992: 200.

34   Garvey 2005: 87, also see account of Pers on p. 97.

35   Collins, Parks, and Marlatt 1985; Borsari and Carey 2001; Sher et al. 2005.

36   Sher et al. 2005.

37   Abrams et al. 2006; Frings et al. 2008

38   Abrams et al. 2006.

39   Elisa Guerra-Doce attributes a marked shift from social to solitary drinking to the Industrial Revolution in Europe and America, citing the work of Schivelbusch 1993 that argues that the rather sudden appearance of counters and bars in drinking establishments serves as a good proxy for the spread of this change. Being forced to drink standing up, or perched alone on an uncomfortable high stool facing the bartender, is a very different experience from sitting around a table with others. Schivelbusch proposes that "the bars sped up drinking, just as the railroad spread up travel and the mechanical loom sped up textile production" (1993: 202, quoted in Guerra-Doce 2020: 69).

40   Earle 2014.

41   "Global status report on alcohol and health 2018" 2018: 261.

42 When it comes to both motor vehicle and non-motor vehicle injuries, "the risk…increases nonlinearly with increasing alcohol consumption"—i.e., not a flat line, but a fairly dramatic curve as one gets to higher levels of BAC (B. Taylor et al. 2010).

43 From "Global status report on alcohol and health 2018," 2018: 89.

44 For a recent review of the problem, and possible remedies, see "Getting to Zero Alcohol-Impaired Driving Fatalities: A Comprehensive Approach to a Persistent Problem" 2018.

45 MacAndrew and Edgerton 1969.

46 Bushman and Cooper 1990; Sher et al. 2005.

47 McKinlay 1951, quoted in Mandelbaum 1965.

48 Lane et al. 2004.

49 Combining figure 1 and figure 2 in Lane et al. 2004, and rounding down the BAC levels to the nearest hundredth of a percent; actual BACs were slightly below .02 percent and above .04 and .08 percent. Original figures copyright © 2003, Springer-Verlag, permission to reuse secured through Copyright Clearance Center, license # 4938450772781.

50 E.g., *Anthony and Cleopatra,* Shakespeare.

51 W. H. George and Stoner 2000.

52 Lee et al. 2008.

53 Archer et al. Under consideration.

54 Abbey, Zawacki, and Buck 2005.

55 Farris, Treat, and Viken 2010.

56 Riemer et al. 2018.

57 See the paper itself for references, which were removed for readability.

58 Barbaree et al. 1983; Norris and Kerr 1993; Markos 2005.

59 See review in Testa et al. 2014.

60 Farris, Treat, and Viken 2010: 427, op cit.

61 Moeran 2005: 26.

62 Yan 2019.

63 T. Wilson 2005: 6.

64 Sowles 2014. Excerpt reprinted with permission.

65 Heath 2000: 164.

66 Testa et al. 2014: 249.

67 Ash Levitt and Cooper 2010; Levitt, Derrick, and Testa 2014.

68 Fairbairn and Testa 2016: 75.

69 Fairbairn and Testa 2016: 74, op cit.

70 *Acts* 2. As Aldous Huxley notes, "It is not only by 'the dry critics of the sober hour' that the state of God-intoxication has been likened to drunkenness. In their efforts to express the inexpressible, the great mystics themselves have done the same. Thus, St. Theresa of Avila tells us that she 'regards the center of our soul as a cellar, into which God admits us as and when it pleases Him, so as to intoxicate us with the delicious wine of His grace'" (*Drugs That Shape,* in Huxley 1954/2009: 8).

71 *Ephesians* 5:18.

72 *Zhuangzi,* Ch. 19; B. Watson 1968: 198-199.

73 Slingerland 2014: Ch. 6.

74   *Heaven and Hell,* in Huxley 1954/2009: 144-145.
75   Newberg et al. 2006.
76   Maurer et al. 1997.
77   See www.stangrof.com/index.php.
78   Vaitl et al. 2005.
79   See the account in Osborne 2014: 196-203.
80   Bloom 1992: 59. Also cf. Frederick Law Olmsted's observations concerning ecstatic black Christian church service (quoted in Ehrenreich 2007: 3) or the phenomenon of "ring shouts" performed by African slaves (Ehrenreich 2007: 127) in the nineteenth-century U.S. It is also worth noting that, during the period when the Camba people studied by Dwight Heath had mostly abandoned their binge-drinking practices, they were not merely newly integrated into peasant collectives, but also evangelical Christianity (Mandelbaum 1965).
81   Wiessner 2014. Thanks for Polly Wiessner for personal communications on this topic. Similarly, the Baku pygmies, another group that eschews alcohol and other chemical intoxicants, has a practice of "nighttime voices," where, in the middle of the night, tribe members can voice controversial or minority views in a consequence-free manner, possibly because being half-asleep creates a hypnagogic state of receptivity (personal communication from Tommy Flint).
82   Raz 2013.
83   Some recent and helpful books include Dean 2017; Warrington 2018; and Willoughby, Tolvi, and Jaeger 2019. There is also a massive academic literature on alcohol-use "harm reduction" pioneered by Alan Marlatt and his students (see, e.g., Larimer and Cronce 2007; Marlatt, Larimer, and Witkiewitz 2012).
84   A. Williams 2019.
85   See review in Bègue et al. 2013.
86   Fromme et al. 1994.
87   On this topic, and for a critique of this view, see Slingerland 2008b.
88   See review in Sher et al. 2005.
89   Davidson 2011, cited in Chrzan 2013: 20.
90   O'Brien 2016, 2018.
91   Quoted in Dean 2017: 24.
92   See, e.g., Berman et al. 2020.
93   Chrzan 2013: 82.
94   Chrzan 2013: 6.
95   Heath 2000: 197.
96   Sowles 2014.
97   Ng Fat, Shelton, and Cable 2018: 1090.
98   Koenig 2019.

CONCLUSION

1   *John* 2:1-11.
2   *Qimin Yaoshu,* quoted in Poo 1999: 134.
3   *Kojiki* no. 49, Miner 1968: 12.
4   Madsen and Madsen 1979: 43.

5   Netting 1964.

6   Quoted in Dietler 2020: 121.

7   T. Wilson 2005: 3. It's worth noting that, like many cultural anthropologists who study alcohol, Wilson declares that "in essence drinking is itself cultural," underplaying the important cross-cultural elements grounded in human biology.

8   "Afterlife," Season 2, Episode 2.

9   Nietzsche 1882/1974: 142. Stuart Walton appropriately uses this quotation as the epigraph for his "Cultural history of intoxication" (2001).

10 The city of Denver also apparently attempted to shut down liquor stores and cannabis dispensaries, an ill-fated effort that lasted less than a day (thanks to Deri Reed for that observation).

11 *The Economist,* "Worth a Shot: A Ban on Sale of Alcohol Begets a Nation of Brewers," April 25, 2020.

12 Manthey et al. 2019.

13 Walton 2000: ix-x.

14 Literally, "not even aware that we have a self."

15 "Drinking Wine," poem #14, from Michael Ing's translation (Ing, in preparation), with significant modifications.

16 www.perseus.tufts.edu/hopper/text?doc=Perseus%3Atext%3A1999.01.0138%3Ahymn%3D7.

17 Osborne 2014: 34.

18 As the classicist Michael Griffin notes (personal communication, Aug. 24, 2020), this ending of the hymn represents a vow from the bard himself, promising to keep Dionysus alive in his memory so that he can continue to, literally, "'cosmify' sweet song," where "cosmify" has the sense of beautifying, setting in good order, adorning or preparing. The idea is that, without drawing inspiration directly from Dionysus as god, the bard's songs will lack a certain beauty or coherence.

INDEX

Page numbers followed by *f* indicate figures; page numbers followed by *t* indicate tables.

abstainers/abstinence. *See also* prohibition
  accommodating social situations for, 279–80
  and alcohol-fueled socializing, 258
  and alcohol in tech industry, 259–60
  among millennials and Generation Z, 280
  Mormons and, 56
  political support for, 53
  in United States, 276–77
  as wrong response to *Lancet* conclusions about alcohol, 217
ABV (alcohol by volume), distilled liquors vs. beer and wine, 232
academic innovation, 169–70
academic research, 179
ACC (anterior cingulate cortex), 99–101
acetaldehyde (C_2H_4O)
  and alcohol metabolism, 46
  and Asian flushing syndrome, 47
  and human metabolism of alcohol, 46*f*
addiction. *See* alcoholism
ADH, 46
ADH4, 43
ADH (alcohol dehydrogenases), 43, 46, 47
administrative hypertrophy, 215
adolescents
  and brain development, 76, 95

and legal drinking age, 278–79
Africa, 284
After Life (TV series), 285
"Against Drinking Wine" (Chinese speech), 51, 52
aggression, 249
agriculture
  alcohol as motivation for, 8
  intoxicant as key to transition from hunter-gatherer to agricultural society, 156
  Neolithic transition as source of stress, 120
  origins of, 107–8
air travel, 123–24
Akin-Amar, 138
Alberti, Marcos, 195, 197, 198
Alcaeus, 117
Alcohol: The World's Favorite Drug (Edwards), 59
alcohol by volume (ABV), distilled liquors vs. beer and wine, 232
alcohol dehydrogenases (ADH), 43, 46, 47
alcohol (generally)
  advantages compared to other intoxicants, 103
  advantages over non-pharmacological ecstatic experiences, 268
  and creativity, 13
  Dionysus and, 105–6
  economic costs and health effects of consumption, 40–41

351

alcohol (*cont.*)
    global adult per capita consumption
      (1990–2017), 288
    hangover theory vs. hijack theory, 4
    and human metabolism, 46–47, 46*f*
    as king of intoxicants, 13–14
    as more social drug than cannabis or
      psychedelics, 190
    origins of term, 235
alcohol use disorder (AUD), 226. *See
    also* alcoholism
alcoholic myopia, 101
alcoholism, 226–31
    estimated global rate of, 226–27, 227*f*
    and genetics, 245
    and Northern drinking culture, 230
    and Southern drinking culture, 229
    U.S. rate of, 227–28
aldehyde dehydrogenase (ALDH), 46,
    47
Allen, Woody, 128
altricial species, 66, 69–70
Amanita muscaria, 150, 323*n*7
American Revolution, 37–38
amygdala, 101, 228
Andaman Islands, 142–43
Andes, 39–40, 113, 142
Andrews, Michael, 167–69
anterior cingulate cortex (ACC), 99–
    101
anthelmintic drugs, 327*n*44
antibacterial qualities of alcohol, 30–31
ants, 65, 88
anxiety reduction, 117–24
aphrodisiac, alcohol as, 193–94
Apollonian–Dionysian conflict, 104–6,
    214–16
applejack, 233
apples, 141
Appleseed, Johnny, 39
Aristotle
    on alcohol and perceived
      attractiveness, 194
    distillation described by, 235
Armenia, 18, 144–45
Army, U.S., 38
Aryans, 9–10

asceticism, 288–89
Asian flushing syndrome, 47–49, 230, 246
asocial learning, 83–84, 84*f*
assassins, 136
attractiveness, perception of, 194–96,
    195*f*, 252
AUD (alcohol use disorder), 226. *See
    also* alcoholism
Australia
    doof, 207
    early fermented beverages in, 19
    kava abuse among indigenous
      population, 242
    pituri, 21, 109–10
    rum during colonial era, 38
Australian Shiraz, 232
availability problems, 46
ayahuasca, 8, 209
Aztecs, 224

Babette's Feast (film), 122–23
babies, learning by, 85
The Bacchae (Greek play), 193
The Bacchanal of the Andrians (Titian), 215
bacchanalia, 22, 104
Bacchus, 22, 214
back, human evolution and, 45–46
bad relationships, reinforcing, 261–64
baijiu, 34
Ballmer, Steve, 164
Ballmer Peak, 164, 164*f*
banquets, 59, 178, 239–40
"Bar Talk: Informal Social Interactions,
    Alcohol Prohibition, and
    Invention" (Andrews), 167–69
bars, 187–92, 275, 339*n*16, 346*n*39
bartenders, base salaries for, 275
Baudelaire, Charles, 221–22
Baum-Baicker, Cynthia, 184
Beamish, Richard, 340*n*38
beer
    in African religious ceremonies, 284
    in ancient Sumer, 38
    and dirty water hypothesis, 33
    and ecstatic ritual, 146
    in *Gilgamesh*, 9
    nutritional value, 326*n*38

in pre-Victorian England, 29
as source of calories, 29
typical ABV, 232
beer before bread hypothesis, 108–10,
  120, 124, 151, 156–58
Beer Goddess, 116
"beer goggles," 252
Bekoff, Marc, 94
Benedikt, Carl, 340*n*40
benefits of intoxication, 217–22
Beowulf, 50
berserkers, 136
binge drinking
  among Vikings, 50
  Camba people and, 262
  decline among millennials, 280
  and Northern drinking culture,
    230
biological ennoblement, 29, 32, 122
biphasal effects of alcohol, 98–99, 225
birds, 330*n*11
blicket, 74, 75*f*
Bloom, Harold, 268
blushing, 126
Bolivia, 136, 261–63
bonding. *See* social bonding
The Book of Mormon, 55–56
Book of Odes, 50–51, 133, 153
Book of Songs, 58
Booth, William, 52–53
Bourguignon, Erika, 142
brain hijack. *See* hijack theories
brain maturation process, 75–76
brandy, 235
Braun, Stephen, 98, 323*n*2
bread, 109
bread before beer hypothesis, 107–8
bridal (term), 134
Brillat-Savarin, Jean Anthelme, xiii, 15
Brin, Sergey, 200
Brodsky, Archie, 191
Bronze Age, 144
Brown, Stuart, 79
Burning Man, 200–201, 207
business travel, 177–82
businessmen, Japanese, 166, 254–55
Byron, Lord, 204

caffeine, 56, 105
Caledonian crow, 66
calories, 29
calumet ("peace pipe"), 132
Camba people, 136, 261–63
Camus, Albert, 206
Canada, 40, 179
Cane Ridge Revival (Kentucky), 268
cannabis, 44, 112
  cognitive effects, 102
  disadvantages of, 103
  earliest evidence of use, 20
  Islam and, 55
  and THC, 43
cannabis dispensaries, 286
Cao Cao (Eastern Han Grand
    Chancellor), 329*n*88
capital riddle, 72
carbon emissions, 180
cassava (manioc), 81–82
causality, learning and, 85
cave tetra, 62–63
"Celebration of the Power of Wine"
    (Liu Ling), 205
Central America, 109
Central Asia, 7
cerebellum, 99
"cerebral reducing valve," 266, 267
Chavín culture, 113
Chengdu, China, 255
chicha, 31, 38–40, 109, 142, 152–53
chickens, 65–66
child mind, regaining, 95–97
children
  cognitive flexibility and creativity, 74,
    78
  and distraction, 69
  and lateral thinking, 115–16
  and learning, 85–86
  and Northern drinking culture, 230
  and PFC development, 96, 115, 145
  and play, 79
  and Southern drinking culture, 229
  trustworthiness detected by, 125
chimpanzees, 61, 83, 128
China
  alcohol consumption in, 34–35

China (*cont.*)
burial practices, 36
business banquets, 255–56
communal drunkenness, 134, 139, 140
and distillation, 235
drinking rituals, 238–40
drunkenness and trust in *Book of Odes*, 133
earliest alcoholic beverages in, 17
evidence against dirty water hypothesis, 34
excess and ecstasy, 205
group labor rewarded by alcohol, 153
and history of prohibition, 50–52
prohibition efforts in, 53
religion in, 58
ritual vessels, 144
Tao Yuanming on power of wine, 220–21
wine and sociability, 187
wine as sacred substance, 284
wine as simultaneous boon and threat, 223
wine definition, 345n1
wine vessels in Shang Dynasty tombs, 38
wine's role in fomenting creativity, 111
wine's role in political agreements, 130
"Written While Drunk" poems, 111
Chinese Communist Party, 223
Christianity, 52–53, 210
chronic stress, 119
Chrzan, Janet, 276–77
Church of the Latter-Day Saints (Mormons), 55–57
ciguatera, 325n13
Clarke, Arthur C., 129
climate change, 180
Clinton, Bill, 128
Code of Hammurabi, 137
coding, BAC and, 164
coffee, 56, 135
cognitive flexibility, 74
Coleridge, Samuel Taylor, 112, 211

collective bonding. *See* social bonding
"collective brain," 80–82
college, drinking at, 274, 277
colonial America, 187, 244
commitment. *See* emotional commitment
communal animals, humans as, 87–95
communal bonding. *See* social bonding
communal drinking, 241
communal drunkenness, 134–41
communal thinking/creativity, 166
Communist Party (China), 223
competitive advantage, 156–58
conflict resolution, 263–64
Confucius, 138–39
contagion effect, 186
cooked food, 63–64
cooperation, 87–95
cooperation challenges, 125
corvids, 67–68
Covid-19 pandemic
and business travel, 177
effect on innovation, 170–71, 340n40
liquor stores as essential service during, 285–86
creative cultures, alcohol in, 260
creative flexibility, narrow competence vs., 69
creativity
anticipated effect of Covid-19 pandemic lockdown on, 180–81
corvids and, 67–68
as human characteristic, 71–79
intoxication as facilitator, 13, 111–16, 162–71
microdosing and, 175
PFC as enemy of, 78
transmitted culture and, 80
crime, distilled liquor and, 237
crowds, alcohol consumption and, 249
crows, 67–68
cultural animals, humans as, 80–87
cultural evolution, 80–84, 246
cultural expectations, 272
cultural group selection, 156–58
cultural innovation, 97, 115, 165, 167–69

cultural learning. *See* social learning
cultural meaning, intoxicants and, 156
cultural norms, alcoholism and, 231
culture
  human dependency on, 63–65,
      84–85
  as key to human survival, 61–64
culture-defining beverages, 284–85
curanderismo, 209
curandero, 209
"curse of the self," 206
Cuzco (Peru), 152–53
Czechoslovakia, 53

Da'an Pan, 111
dance, 142–43, 201, 202
Dancing in the Streets (Ehrenreich), 207
Daodejing (Laozi), 97
Daoism, 97, 265, 269–70
Darwin, Charles, 126
date rape drug, alcohol as, 251
Davidson, James, 273
Dean, Rosamund, 273
death
  alcohol consumption and, 247–51,
      247*t*–48*t*
  alcohol-related (2016), 225–26
deception, 128–29
default-mode network (DMN), 102
defection
  and blushing, 126
  commitment relationships and, 124
  Odysseus and, 92
  primates and, 88–89
  Prisoner's Dilemma and, 90–91, 125
  sobriety and, 132
delayed gratification, 78, 96
Deng Xiaoping, 133
depressant, alcohol as, 99, 225, 236
dian, 58
Dietary Guidelines for Americans, 160
Dietler, Michael, 324*n*10
Dietrich, Arne, 102
Dietrich, Oliver, 151
Dionysian experience, 204–5
Dionysian–Apollonian conflict, 104–6,
    214–16

Dionysus
  and bacchanalia, 22
  and danger, 270
  earliest myth about, 290–91
  as founder of civilization, 110
  as god, 291
  as god of chaos and disorder, 223
  Nietzsche and, 204
  on virtues of moderation, 52
dirty water hypothesis, 32–35
dishonesty, detecting, 127–31
disinfectant, alcohol as, 30–31
disinhibition, 178–79
Disraeli, Benjamin, 199
dissociation, 142
distillation process, 234–38
distilled liquors
  as evolutionary mismatch, 232–38
  introduction to South American
      cultures, 244–45
  isolation and, 244–46
  and Northern drinking culture, 230
  raising drinking age for, 279
  in Russia, 245
  as separate drug from other alcoholic
      beverages, 274
  shipwreck survivors and, 330*n*2
  and Southern drinking culture, 229
distraction tasks, 162–63
DMN (default-mode network), 102
DNA, 87
dogs, 78–79
Doniger, Wendy, 323*n*7
doof, 207
The Doors of Perception (Huxley), 106
dopamine, 101, 102, 194, 196
dopamine receptors, 228
Doughty, Paul, 153
drinking age, 278–79
drinking alone. *See* solitary drinking
drug instrumentalization, 217
drumming, 267
drunken mind, 97–103
drunken monkey hypothesis, 27–28
dualism, mind-body, 218, 220, 221
Duchenne smiles, 185, 185*f*
Dudley, Robert, 27–28

Duke, Guy, 142, 152–53
Dunbar, Robin
    on alcohol as social drug, 190
    on alcohol's role in social rituals,
        147–48
    on biotechnology of group
        formation, 201
    on pubs, 189, 190
Durkheim, Émile
    on collective effervescence, 143
    on ecstatic experience, 202, 203

Earle, Rebecca, 244
East Asians, 47, 49
Eastern Europe, alcoholism in, 231
Eastern Han Dynasty, 329n88
eboka, 146
ecological niche, 62–70, 104–6
economic commodities, alcoholic
        beverages as, 338n125
economic costs of alcoholism, 227
ecstatic experiences, 202–11
    bonding and, 199–203
    Burning Man, 200–201
    definition and origins of term, 142
    religion and, 266
    and transformation, 150
ecstatic rituals, 207
Edwards, Griffith, 59, 189
Egypt, ancient
    alcoholism in, 226
    beer in mythology, 39
    Festival of Drunkenness, 143–44
    wine vessels in tombs, 38
Ehrenreich, Barbara, 201, 207, 210
Eisbock beers, 233
elephants, alcohol and, 328n68
Eliade, Mircea, 201, 219
Elizabethan England, 131
Emerson, Ralph Waldo, 141
emotional commitment, 90–93, 124. See
        also trust-based relationships
emotional leakage, 126
endorphins, 3, 99, 148, 190
Engels, Friedrich, 166
England
    "gin craze" in 1700s, 237
    pubs in, 187–91
    vows and toasting, 131
Enkidu, 9, 192
Enright, Michael, 154–55
enzymes, 43
Ephesians, 265
Erlitou culture, 152
espionage, 179
ethanol (C2H6O). See also alcohol
        (generally)
    advantages as intoxicant, 285
    and alcohol metabolism, 46
    complexity of inebriation effect,
        98–99
    and distillation, 234–36
    and fractional freezing, 232–33
    and human metabolism, 46f
    and psychological/behavioral effects
        of alcohol, 272
    in sacred or culture-defining
        beverages, 285
    in vodka, 236
Europe, distilled liquors in, 235
evolution
    and alcohol dependence, 228, 229
    appearance of distilled liquor on
        evolutionary time scale, 237–38,
        238f
    cultural, 80–84
    and distilled liquor, 245
    distilled liquors as evolutionary
        mismatch, 232–38
    hijacking/gaming of system, 3
    intoxicants and, 10, 11
    intoxication as evolutionary puzzle,
        35–41
evolutionary hijacking. See hijack
        theories
evolutionary mistakes, 4–5
expectancy effects, 121, 271, 272
experiments
    alcohol as facilitator of social
        bonding, 184–86
    distraction tasks, 162–63
    stress in humans, 120–22
    stress in rats, 118–20

facial expressions, 126, 185, 185*f*
Fang culture, 146
fermentation
  bread before beer hypothesis, 107–8
  earliest evidence of, 109
  in human history, 39–40
  nutrition and, 28–29
Ferriss, Tim, 174–75
Fertile Crescent, 107, 109, 120
Festival of Drunkenness, 143–44
festivals, 203
festive labor, 153
Fiji, 132, 140, 155
fire, 63–64
fitness benefit, 25
flushing reaction, 47–49
fly agaric toadstool, 150
food preservation, evolutionary
  hangover theories and, 32
football, 249
Forsyth, Mark, 39, 50, 144
fortune-tellers, 129
Foster, Georgia, 275
fractional freezing, 232–34
Frank, Robert, 92
frat parties, 154
fraternity hazing rituals, 243
Frederick the Great (king of Prussia),
  135
free-rider problem, 89
French paradox, 159–60
Friedman, Kinky, 194
friendship, 197–98
frontal cortex, development of, 76–78.
  See also prefrontal cortex (PFC)
fruit flies *(Drosophila)*, 26–27, 30, 252
Fuller, Robert
  on calumet, 132
  on ecstatic ritual, 146
  on Mormon ban on psychoactive
    chemicals, 57
  on peyote ceremony, 149
fun, intoxication as, 211–16
Furst, Peter, 149

GABA, 228
$GABA_A$ receptors, 99

Gabon, 146
Game of Thrones (TV program), 131–32,
  258–59
Gately, Iain
  on binge drinking in Viking culture,
    50
  on decision-making in ancient Persia,
    166
  on Kvasir, 112
  on sobriety, 131
Gay Hussar restaurant (Soho), 190, 257
gender/gender roles
  and after-work drinking, 256–57
  and alcohol-fueled social bonding,
    136
  and drinking, 136
  and Iron Age burial practices, 154–55
  and old boys' clubs, 260–61
gene-culture coevolution, 324n10
general intelligence, 69
Generation Z, 280
Genesis (biblical book), 251
genetic evolution, 41–49
genetics, alcohol dependence and, 228,
  229, 245
Georgia, Republic of, 17, 241
Germany, 137
Gervais, Ricky, 285
"Get Drunk" (Baudelaire), 221–22
Getaway (Brooklyn sober bar), 271
Ghost Dance, 149–50
Gilgamesh, 8–9
gin, 236
glossolalia, 267
Göbekli Tepe, 108, 145, 150–52, 150*f,*
  153, 156
Good Friday Experiment, 208
Google
  alcohol's role in institutional life, 183
  and Burning Man, 200
  whiskey rooms, 164–65
Gopnik, Alison
  on children's creativity, 96
  on cognitive/creative flexibility, 69,
    74
  on learning, 85
government policy, 159

grape domestication, earliest, 17–18
grape wine production, earliest, 144–45
gratification, delayed, 78, 96
gratuities, servers' dependence on, 275
gray matter, 75–78
Great Britain
  "gin craze" in 1700s, 237
  pubs in, 187–91
Greece, ancient
  alcohol consumption at sporting
    events, 249
  communal drunkenness, 139, 140
  contempt for drunkards, 52
  host's responsibility for moderating
    drinking at symposia, 241
  methods for encouraging
    moderation in drinking, 273
  sex and Dionysian rituals, 193
  symposiarchs, 139, 274
  symposium, 141
  and undiluted wine, 235
  virtue and intoxication, 215–16
  "water drinkers" in, 291
Greek mythology, 92. See also
  Apollonian–Dionysian conflict
Griffin, Michael, 349n18
Grof, Stanislav, 268
Guerra-Doce, Elisa, 346n39
Guinness, 29

Hafez of Shiraz, 55
Hagen, Ed, 43
Haidt, Jonathan, 201–2
hallucinogens. See psychedelics
Halsrätsel (capital riddle), 72
hangover theories, 3, 27–31
  and alcoholism, 230
  and alcohol's antibacterial qualities,
    30
  defined, 27
  and distilled liquor, 245
  Dudley's drunken monkey theory,
    27–28
  and human predilection for
    intoxication, 44–45
  weakness of theory, 31–32
Hart, H. H., 192

Harvey, Larry, 200, 201
hashish, 55, 136
Hathor (goddess), 39, 143–44
Hayden, Brian, 120, 122
hazing rituals, 243
HDL (high-density lipoprotein), 160
Head of Table (tamada), 241
health consequences, of alcohol
  consumption, 247–51, 247t–48t
heart disease, French paradox and, 160
Heath, Dwight
  on bonding over drink, 134
  on drinking among Camba people,
    136
  on drinking as source of pleasure,
    214
  fieldwork on Camba people, 261–63
  on societal desire to regulate alcohol,
    224–25
  on societies in which alcohol isn't
    well integrated into social life,
    277
  on solitary drinking, 241
hedonism, 211–22
Heilig, Markus, 228
Hemingway, Ernest, 112
Henrich, Joseph, 81, 82, 325n13
Herodotus, 20
Hessians, 37, 135
hierarchies, 94, 155, 166, 254, 261
hijack theories, 24–27
  alcohol and, 4, 11
  and alcoholism, 230
  defined, 4
  overlap with hangover theories, 27
  and plant-based recreational drugs,
    43
  weakness of, 31–32
hippocampus, 99
hive mind, 123, 141–50, 147, 202
Hockings, Kimberley, 190
Hogan, Emma, 174–75, 201
Hölderlin, Friedrich, 111
"holotropic breathwork," 268
home, drinking at. See solitary
  drinking
homeostasis, 228

Homo neanderthalensis, 113
Horton, Donald, 117–18
HRAF (Human Relations Area File) project, 146
Hrdy, Sarah Blaffer, 64
Hubbard, Al, 174
Huizinga, Johan, 72, 73, 78
human ecological niche. *See* ecological niche
Human Relations Area File (HRAF) project, 146
hunter-gatherers, 120, 156
Huxley, Aldous
    on chemically induced spiritual experience, 172–73, 219–20
    on need for intoxication, 106
    on need to escape from self, 206
    on religious alternatives to chemical intoxication, 266–67
hydraulic model of alcohol use, 118
hyperventilation, 268
hypocrisy and hypocrites, 127

"I Drink Alone" (Thorogood), 241
Ibn Fadlan, Ahmad, 49–50, 54
in vino veritas, 131
Inca Empire, 38, 109, 143, 152–53
inclusivity, alcohol in tech industry and, 259–60
India, 19
Indiana University, 121
individualism, 210, 245
Industrial Workers of the World (IWW; Wobblies), 134–35
industrialized societies
    decline of ritual and ecstatic bonding in, 202
    drinking by married couples in, 263–64
    drinking on the job, 153
    drinking sessions in Japan, 166
    gender roles in, 257
    microenvironments in, 218
    non-drinkers in, 280
Ing, Michael, 197, 205
injury, alcohol consumption and, 247*t*–48*t*

innovation
    anticipated effect of Covid-19 pandemic lockdown on, 180–81
    childlike state of mind and, 96–97
    and cultural evolution, 82
    and lateral thinking, 115
    and socializing at academic conferences, 180
insider groups, 254–61
internet
    individualism and, 211
    videoconferencing, 170–71, 177, 180–82
intimacy, 186, 192–98
intoxicants. *See also specific intoxicants*
    in cultures that do not produce alcohol, 7–8
    as evolutionary mystery, 6
    social benefits of, 286–88
    uses in industrialized societies, 218
intoxication (generally)
    and adaptation to ecological niche, 62
    as antidote to PFC, 104
    broad use of term, 6–7
    and creativity, 111–16
    in early human history, 17–28
    as evolutionary puzzle, 35–41
    negative consequences of, 223–82
    non-chemical alternatives, 264–70
    and origins of agriculture, 110
    pharmacological vs. non-pharmacological, 268–69
    prejudice against, 218–19
    reasons for, 17–60
    as solution for various human challenges, 12
    as taboo topic, 289
    as vice, 5–6
introverts, 198
iPhone, 80
Iran, 18
Iron Age, 154–55
Islam, 48, 54–55, 57
isolation. *See* solitary drinking
Italy, 229, 231
IWW (Industrial Workers of the World; "Wobblies"), 134–35

James, William, 141, 172
Japan
    alcohol consumption during Shinto
        rituals, 241
    and sake, 284
    salarymen's drinking sessions, 166,
        254–55
Jarosz, Andrew, 162–63
jazz improvisation, 78
Jennings, Justin, 136, 137
Jesus Christ, 283–84
Jews and Judaism, 133
Jiahu tomb, 144
jiu (wine), 58
Jobs, Steve, 174
Johnson, Hugh, 213
Jordan, 109
junk food, as evolutionary hangover, 4,
    5, 27, 41

kaishui, 34
kanikani, 140
kava
    abuse of, 242
    cognitive effects, 102
    and creativity, 112
    health effects of, 40
    origins, 20
    and social prestige, 140, 155
    and truth-telling, 132
khamr, 54–55
Khazan, Olga, 160–61
Kissinger, Henry, 133
Kofyar people (Nigeria), 284
Koori people (Australia), 19
Kotler, Steven, 199
krater, 154
"Kubla Khan" (Coleridge), 112, 211
!Kung people, 268
Kvasir (Norse god), 111–12
Kwong, Charles, 220–21

Lalander, Philip, 166
The Lancet, 160–61, 212, 217, 288
lateral thinking, 73–74
    and brain development, 78
    intoxication and, 114–16

PFC and, 78
    reduced cognitive control and, 162,
        163
    learning, PFC and, 86–87
Leary, Mark, 206
Leary, Timothy, 176
lie detection, 178
liquor stores, as essential service during
    Covid-19 lockdown, 286
literature, alcohol and, 112
Liu Ling, 205
"local" (pub), 187
London, Jack, 167
"Lost in the Sauce" (Sayette, et al.),
    163–64
The Lost Weekend (film), 112
Lot (biblical figure), 251
loyalty, 90–91
LSD, 174–76
lying, 126, 178

mada, 19
Madsen, William, 137
Magellan, Ferdinand, 327*n*49
maize beer, 326*n*38
manioc, 81–82
Marcinko, Richard, 199
Mardi Gras, 207
married couples, effect of drinking
    together, 263–64
Mars, Gerald, 140
Marshall, Alfred, 167
Marx, Karl, 11–12, 166
Masaryk, Tomáš, 53, 54
masturbation, 3–5, 24
matched drinking, 242
Matthew, Gospel of, 97
Mattice, Sarah, 139
Maximator, 179
Mayan culture, 143
McGovern, Patrick, 8
MDMA, 148, 207–8
mead hall, 187
meals, drinking at, 229, 276, 278
menopause, 70
mescal bean, 22–23
mescaline, 113, 149, 172–73

Mesopotamia, 29, 134, 192–93
mestizo, 241
metabolism of alcohol, 46–47, 46*f*
Mexico, 7, 19, 137, 148–49, 284
Michalowski, Piotr, 138, 193
microdosing, 174–75
microenvironments, 218
Microsoft, 164
Midas, King, 110
Middle Ages, 235
millennials, 280
"Mind-at-Large," 266, 267
mind-body dualism, 218, 220, 221
mind wandering, 114
mindful drinking, 273–74
Mindful Drinking (Dean), 273
moderate drinkers, 191, 197, 198
moderate drinking, 211–12, 278
modern world, intoxication in,
    159–22
moralism, in contemporary era, 289
Mormons (Church of Latter Day
    Saints), 55–57
Morris, Steve, 328*n*68
mortality
  alcohol consumption and, 247–51,
    247*t*–48*t*
  alcohol-related (2016), 225–26
Mott, Edward Spencer, 59
Müller, Christian, 197, 217
Munroe, Randall, 165
music, 142–43, 146
music festivals, 203
Muthukrishna, Michael, 80, 83
myopia, 101

narrow competence, creative flexibility
    vs., 69
National Institutes of Health, 226
Native Americans, 244
natural selection, 4–5
Navy Seals, 199
NCL (nidopallium caudolaterale),
    330*n*11
Neanderthals, 113
negative consequences of intoxication,
    223–82

negative feedback
  AAC and, 99–100
  alcohol and, 251
  alcohol consumption and insensitivity
    to, 251
  Wisconsin Card Sorting Test and, 101
negative selection, 230
neo-Prohibitionists, 160–61
Neolithic, 17–19, 120, 144, 145
neoteny, 79
Nesse, Randolph, 25
neural pruning, 75–76
neurotoxins, 22, 327*n*44. *See also specific*
    neurotoxins
Nevali Çori, 145*f*
"new sobriety," 281
New South Wales, Australia, 38
New Testament, 264–65
Newfoundland, 140
niche environments, 62
Nicotiana attenuata, 325*n*18
Nicotiana rustica, 325*n*18
nicotine, 105
nidopallium caudolaterale (NCL),
    330*n*11
Nietzsche, Friedrich, 204, 285
Nigeria, 284
nitrous oxide, 172
Noah, 224
non-alcoholic drinks, 279–80, 287. *See*
    also virgin cocktails
Norse mythology, 111–12, 136
Northern drinking culture, 230, 231,
    276–78
Northern Europe, alcoholism in, 231
Norway, 242, 276
nunc est bidendum ("Now is the time for
    drinking"), 169
nutritional value of beer, 326*n*38
Nutt, David, 175

The Oasis (bar), 339*n*16
objectification of women, 252–53
Odin (Norse god), 50
Odysseus, 92
Oedipus, 71–72
Oedipus Rex (Sophocles), 71–72

Office Christmas Party (film), 183
office parties, 182–87
old boys' clubs, 254–61
O'Neill, Eugene, 112
opium, 21
"oracle bones," 58
orgasms, 4, 23–24
Orkney Islands, 18
Ortega y Gasset, José, 214
Osborne, Robin, 342n65
Otto, Adelheid, 29
Out of It (Walton), 212–13
Oxford University, 169

Pacific Islands, 20
Page, Larry, 200
palm wine, 155
Paris, France, 166–67
passing out, BAC and, 236–37
patent registration, Prohibition's effect on, 168–69, 256–57
path dependence, 46
Paul (apostle), 265
"peace pipe" (calumet), 132
Peele, Stanton, 191
Pentecostal ecstatic rituals, 147, 267
Persia, 166, 235
Peru, 137, 153, 209, 241
Peter (apostle), 265, 267
peyote, 19, 135–36, 148–49
PFC. *See* prefrontal cortex
Philostratus, 196
phonotelephote, 177
physical aggression, 249
Picturephone, 177
Pinker, Steven, 26
Pitek, Emily, 146
pituri, 21, 109–10
placebo effect, 121, 122, 271
Plains tribes, 132
Plato, 106, 131, 139
Platt, B. S., 29
play, 78–79, 85–86, 93–95
playfulness, 193
politics/political power, 150–55, 190–91
Pollan, Michael, 39, 176, 323n2
Poo Mu-chou, 58, 239

pornography, 23–26, 41
porridge, 32
pre-commitment, 92
precocial species, 66, 68
prefrontal cortex (PFC)
    adulthood and, 95–96
    alcohol as means of deactivating, 97–103, 105
    and alcohol-fueled socializing, 258
    as Apollonian, 104–5
    and brain development, 76, 78
    and cognitive control tasks, 163
    as creativity inhibitor, 114–15
    deactivating, 102
    intoxicants as essential tools for living with, 285
    and learning, 86–87
    and Prisoner's Dilemma, 91
    pros and cons of, 103–4
    psychedelics' effect on, 173
    and religious intoxication, 265, 267
    and role of drinking in business negotiations, 178
    and Southern drinking culture, 278
    as target of intoxication, 6, 7
    and trustworthiness, 125, 133
    and Wisconsin Card Sorting Test, 100–101
primates
    adaptation to ecological niche, 63
    and selfishness, 88–89
Prisoner's Dilemma, 89–91, 90t
    2020 Democratic primary and, 332n45
    serotonin and, 148
    trust and, 93, 124, 125
problem drinking, 277
prohibition
    in China, 51, 53, 329n88
    decrease in new inventions during, 168–69, 256–57
    failure of, 49–57
    Islam and, 54–55
    Mormons and, 56, 57
    Sri Lanka's attempt to impose, 286
    in U.S., 167–69, 234
Proverbs (biblical book), 117

psilocybin, 102, 103, 208
psychedelics (hallucinogens), 171–77
  berserkers and, 136
  cognitive effects, 102
  and creativity, 112–13
  dangerous chemical makeup of,
      22–23
  disadvantages of, 103
  in early beers, 19
  early human use of, 19–20
  and ecstatic ritual, 146, 148–49,
      207–8
  microdosing, 174–75
  safety relative to alcohol or cannabis,
      175
  and Silicon Valley innovations,
      173–74
  as treatment for mental disorders,
      208–9
psychopaths, 128–29
The Pub and the People, 187–89
public drunkenness, 230
public health, 159–60
pubs, 169, 187–92, 256, 257
pulque, 137, 224, 284
Purim, 133

qi, 34
Qin, First Emperor of, 36

Ra (Egyptian god), 39
Radcliffe-Brown, Alfred, 142
Rappaport, Roy, 143, 202
RAT (Remote Associates Test), 73, 114,
    162
rats, stress studies with, 118–20, 122
Raz, Gil, 269–70
recreational cannabis dispensaries, 286
red states, abstinence in, 277
religion, 58–59
  as evolutionary puzzle, 35–37
  non-chemical alternatives to
      intoxication, 264–70
  sacred alcoholic beverages, 284–85
  speaking in tongues, 264–65
Remote Associates Test (RAT), 73, 114,
    162

"Returning Home" poems (Tao
    Yuanming), 220
Revivalist Christianity, 268
reward systems, 25
riddles, 71–72
Rig Veda, 9–10
risk-taking, 249–51, 250f
ritual, ecstatic, 141–50
Ritual and Religion in the Making of
    Humanity (Rappaport), 202
Roberts, Richard, 168
Rogers, Adam, 192, 234
Roth, Marty
  on alcohol and creativity, 112
  on drinking as source of pleasure,
      214
  on intoxication and religious ecstasy,
      219
rum, 236
runner's high, 102
Russia
  fractured social order and distilled
      liquors in, 245
  and Northern drinking culture, 230
  vodka consumption after fall of
      Communism, 237, 244

sake, 284
salaries, for servers/bartenders, 275
Salvation Army, 52–53
Samorini, Giorgio, 176
San Francisco, 174
San Pedro cactus, 113
Sayette, Michael, 163–64, 184–86,
    336n60
Schmidt, Eric, 200
Schumann, Gunter, 197, 217
Schwips, 236
Scythians, 20
SEAL Team Six, 199
self-awareness, 196, 206, 211, 285
self-inflation effect, 196
selfishness, primates and, 88–89
selflessness, 141–50
serotonin, 99, 101, 102, 148,
    150
servers, base salaries for, 274–75

sex
alcohol and, 192–98
consensual vs. nonconsensual, 253
fruit flies and, 26–27
masturbation as evolutionary hijack
of impulse, 3–5
and pleasure, 26–27
pornography and, 23–24
violence against women, 251–54
sexual harassment, 257, 261
sexual violence, 251–54
Shakespeare, William, 194
"shamanic-type" experiences, 267–68
shamanism, 219
shamans, 113–14
Shang Dynasty, 38, 51, 58, 152
shaojiu (cooked/distilled wine), 235
Sharon, Douglas, 209
Shaver, John, 140
Sherratt, Andrew, 21–22
Shinto, 241
shipwreck survivors, 330*n*2
Shiraz, 232
Shulgin, Alexander, 207–8
Siegel, Ronald, 206, 328*n*67
Sierra Madre Occidental mountain
range, 148–49
Silicon Valley. *See also* tech industry
alcohol-fueled decision-making in,
166
Burning Man and, 200
psychedelics' effect on innovation,
173–74
Sima Qian, 240
sindicatos, 262, 263
sirens, 92
Skype, 177
sleep aid, alcohol's deficiencies as,
332*n*58
Smail, Daniel, 235–36
smiling, 185, 185*f*, 186
Smith, Joseph, 55–56
sober bars, 270–73
"sober curious," 271
sobriety, office parties and, 183
soccer, 249
social bonding

and after-work drinking, 257–58
and alcohol in tech industry, 259–60
and Camba people, 261–63
children and, 93
communal drunkenness and, 134–41
ecstatic experiences and, 199–203
experiments with alcohol as
facilitator of, 184–86
gender and, 136
intoxication as facilitator, 12
non-alcoholic vs. alcohol-assisted, 287
office parties and, 183–84
and old boys' clubs, 254–61
psychedelics and, 148–50
pubs' role in, 189–92
warriors and, 135–36
social cohesion, 156–58, 186
social drinking, 238–41
drinking in isolation vs., 243
liquor and, 236
Prohibition and, 168
pubs and, 187–92
social emotions, 91, 92
social hierarchy, 94, 155, 166, 254
social insects, 65, 88
social learning, 83–84, 84*f*, 85
social monitoring, 242
social networks, as collective brains, 80
social phobias, 198
social regulation, 240–41
social skills, 61–64
social solidarity, 134–41, 150–55, 262–
63
social status, 154
social toasts, 59
socializing, academic conferences and,
180
Socrates, 139, 193
software coding, 164–65
solidarity, 134–41, 150–55, 262–63
solitary drinking, 238–44
and alcoholism, 231
bars and, 346*n*39
dangers of, 275–76
distilled liquors and, 244–46
and epidemics of alcohol abuse, 244–
45

and risky behavior, 243
as symptom of problem drinking, 241
soma, 9–10, 19, 150
song, ecstatic union and, 142–43
Song of Songs (biblical book), 193
Sophora secundiflora, 22
Sosis, Richard, 140
South America, 109
South Carolina, 277
Southern Europe, 229
Southern (European) drinking
cultures, 229, 276
means of instilling young adults into,
278–79
as model for adapting to distillation/
isolation problem, 246
as model for normalizing alcohol
consumption, 276, 278–79
Soviet Union, 237, 244
Sowles, Kara, 259–60, 279–80
Spain, 231
speaking in tongues, 265, 267
Sphinx, 71–72
Spiegelhalter, Sir David, 217
spies, 179
spine, human evolution and, 45–46
sporting events, alcohol consumption
at, 249
Sri Lanka, 286
Stalin, Joseph, 336*n*60
Stealing Fire (Kotler and Wheal), 199
stimulants, 6, 98–99, 225
stress, 118–20, 119*f*
stress reduction, 116–24
stress-response dampening, 122
suburban communities, isolation in,
243–45
Sufism, 55, 268
Sumer/Sumerians
beer in poetry of, 193
betrayal of oaths made when drunk,
138
drinking ceremonies, 155
early beer recipes, 109
Enkidu myth, 39, 192
grain production devoted to beer, 38
hymn to Beer Goddess, 116

myths of origins of civilization, 8–9
and origins of beer, 109
unfiltered beer, 333*n*4
Survivor (reality TV show), 61
Sweden, 166, 276
symposiarch, 139, 274–75
symposium
host's responsibility for moderating
drinking at, 241
intoxication at, 141
origin of term, 167
as test of character, 139
Symposium (Plato), 131, 193
Szaif, Jan, 215–16

Tacitus, 130–31
tamada (Head of Table), 241
Tang Dynasty, 235
Tanzania, 29, 284
Tao Yuanming
on friendship, 197–98
on human need for wine, 117
on the power of wine, 220–21
on wine as source of meaning and
pleasure, 290
Tasmania, 19
taverns, 187
tea, 34, 56
tech industry
alcohol-fueled decision-making in,
166
alcohol in, 259–60
Burning Man and, 200
psychedelics' effect on, 173–74, 201
teenagers
and brain development, 76, 95
and legal drinking age, 278–79
teleconferencing. *See*
videoconferencing
ten Brinke, Leanne, 126
teosinte, 109
tetra, 62–63
THC, 43, 44, 102
Thorogood, George, 241
Three Cs (creative, cultural,
communal), 70, 95–97
tips, servers' dependence on, 275

Titian, 214, 215
toads, poisonous, 8, 19–20
toast leader (symposiarch), 139, 274
toasts, 59
tobacco, 21, 56, 110
Tohono O'odham people, 240
Tolstoy, Leo, 116
tombs
 Andes, 113
 cannabis in early Central Eurasia, 20
 First Emperor of Qin, 36
 Iron Age Europe, 154–55
 Jiahu, 144
 Shang Dynasty, 38, 156
 Zhou Dynasty, 53
tools, cultural evolution and, 82
toxins, defined, 5–6
traditional healing, psychedelics and, 208–10
tragedy of the commons, 89
transcranial magnet, 78, 97–98, 162
transformation and transformative experiences
 alcohol as drug least conducive to, 343n98
 Burning Man and, 200
 Dionysus and, 110
 drugs and, 203
 ecstatic experience and, 150
 in Persian poetry, 219
"transparent" drugs, 323n2
trust, 64, 93, 132–34
trust-based relationships, 124–33
trustworthiness, evaluation of, 124–28
truth-telling, alcohol and, 130–32
tuberculosis, 47
Turkey, 108, 145
"turn on, tune in, drop out," 176–77
Turner, Fred, 200
Twain, Mark, 39
Twinkies, as evolutionary hangover, 4

unconsciousness, 136, 236–37
"Uncorking the Muse: Alcohol intoxication facilitates creative problem solving" (Jarosz, et al.), 162–63

United Kingdom, pubs in, 187–90
United States
 alcoholism rate, 227–28
 economic costs and health effects of alcohol consumption, 40
 and Northern drinking culture, 276–77
 solitary drinking in, 241
unity, 141–50
University of British Columbia, 169
Unusual Uses Task (UUT), 73, 175
U.S. Army, 38
U.S. Navy Seals, 199

"vacation from the self," 204–11
Vaitl, Dieter, 268
Vanuatu, 20, 112
vapor, distillation and, 234
Vedic people, 19, 150
Velázquez, Diego, 214
"Venus with a horn from Laussel," 18f
Verne, Jules, 177
vice, defined, 5
videoconferencing, 170–71, 177, 180–82
Vikings, 49–50, 53–54, 131
violence, alcohol and, 249
violence against women, 251–54
virgin cocktails, 121, 163, 184–86, 271–72
virtual meetings, 170–71. See also videoconferencing
vodka, 236, 237, 244
Volga Bulgars, 49–50

Wadley, Greg, 120, 122
Walton, Stuart
 on intoxication as escape from moderation, 204
 on intoxication as its own justification, 215
 on intoxication as taboo topic, 212–13, 289
 on post-Temperance Christianity, 210
War and Peace (Tolstoy), 163–64
Warrington, Ruby, 270–71
warriors, bonding among, 135–36

Washington, George, 37–38, 135
Wasson, Gordon, 150, 323n7
water. *See also* dirty water hypothesis
    access to, in social situations, 279–80
    disinfectant effect of fermentation,
        30–31
    and distillation, 234
    and fractional freezing, 232–33
"water drinkers," 291
"water trade," 254
weddings, 134
Weil, Andrew, 21
Western Zhou Dynasty, 51
Wheal, Jamie, 199
wheat beer, 326n38
whirling dervishes, 268
whiskey, 39. *See also* distilled liquors
whiskey rooms, 164–65
white matter, 75–78
Wilder, Billy, 112
Williams, Nicholas, 345n1
Wilson, Thomas, 284
wine
    and brandy, 235
    Chinese use of term, 345n1
    Dionysus as god of, 105
    Jesus' conversion of water into,
        283–84
    origins, 17–18
    as sacred substance in medieval
        China, 284
    in Shang Dynasty, 58
    and Southern drinking culture,
        229
    Tao Yuanming on power of, 220–21
    typical ABV, 232
Wisconsin Card Sorting Test, 100–101
Wobblies (Industrial Workers of the
    World), 134–35

women
    and after-work drinking, 256–57
    alcohol access in ancient Greece,
        154–55
    and Chinese business banquets,
        255–56
    cross-cultural patterns of drinking,
        136
    exclusion from old boys' clubs,
        260–61
    and Japanese business culture,
        254–55
    Prohibition's effect on patent
        applications by, 257
    sexuality and alcohol, 194
    and solitary drinking, 275
    violence against, 251–54
Woodlands tribes, 132
Wooldridge, Adrian, 190–91
"Word of Wisdom," 56
World Health Organization, 225–26
Wrangham, Richard, 63

Xia Dynasty, 51

Yan Ge, 255–56
yeast, alcohol tolerance of, 27–28, 30,
    232
yeast-bacteria warfare, 28
Yellow River Valley (China), 17, 144,
    152
Yu (Xia Dynasty leader), 51

Zhang Yue, 111
Zhou Dynasty, 52
Zhuangzi, 265–66
Zoom meetings, 170–71

ABOUT THE AUTHOR

Edward Slingerland is Distinguished University Scholar and Professor of Philosophy at the University of British Columbia, with adjunct appointments in Asian Studies and Psychology, as well as Co-director of the Centre for the Study of Human Evolution, Cognition and Culture and Director of the Database of Religious History (DRH). Slingerland is the author of *Trying Not to Try*, which was named one of the best books of 2014 by *The Guardian* and *Brain Pickings*, as well as academic monographs, translations, and edited books and articles in top journals in a variety of fields. He has given talks on the science and power of spontaneity at venues across the world, including TEDx Maastricht and two Google campuses, and has done numerous interviews on TV, radio, blogs, and podcasts, including NPR, the BBC World Service, and the CBC.

edwardslingerland.com